Statistics for nuclear and particle physicists

LOUIS LYONS

Department of Nuclear Physics
University of Oxford

CAMBRIDGE
UNIVERSITY PRESS

Published by the Press Syndicate of the University of Cambridge
The Pitt Building, Trumpington Street, Cambridge CB2 1RP
40 West 20th Street, New York, NY 10011–4211, USA
10 Stamford Road, Oakleigh, Victoria 3166, Australia

First published 1986
First paperback edition 1989
Reprinted 1992

Printed in Great Britain at the University Press, Cambridge

British Library cataloguing in publication data

Lyons, Louis
Statistics for nuclear and particle physicists.
1. Quantum theory—Statistical methods   I. Title
530.1'2   QC174.17.S7

Library of Congress cataloguing in publication data

Lyons, Louis
Statistics for nuclear and particle physicists.

Bibliography; p.
Includes index.
1. Nuclear physics—Statistical methods.
2. Particles (Nuclear physics)—Statistical methods.
3. Statistics.   I. Title.
QC776.L96   1986   539.7'01'5195   85-11041

ISBN 0 521 25540 6 hardback
ISBN 0 521 37934 2 paperback

UP

Statistics for nuclear and
particle physicists

'The result of this experiment was inconclusive,
so we had to use statistics.'

(Overheard at international physics conference)

CONTENTS

PREFACE

The material in this book is an enlarged version of a highly successful, short course of lectures given to graduate students in the Nuclear Physics Laboratory, Oxford. The course was designed to interest both nuclear structure and elementary particle physicists.

I am an experimental high energy physicist, and although as an undergraduate I suffered a course in statistics, it was only later during my research work, when I had to deduce something from my own data, that I learned how to use statistics. I also discovered that there were many things that one learned by experience and which were not explicitly mentioned in text-books; I tried to incorporate these aspects of the subject in my lectures. The emphasis of my lectures was thus not on formal proofs or on a rigorous treatment of the subject, but rather on practical applications and on how to use statistics to obtain the best results from one's data and to know the limitations of the results. In short, this was a course given by a non-statistician to non-statisticians.

My course did not attempt to cover every single example of statistics problems that can arise in nuclear and particle physics. The aim was rather to explain as fully as possible the different techniques that are available for attacking data analysis problems, to explain their relative merits and drawbacks, and to try to give the students sufficient confidence in their own ability to tackle any new problems that they might encounter. The book has maintained this approach.

There is an old joke about a man who was asked whether he could play the violin, to which he replied that he did not know as he had never tried. Instead of being about violin-playing, this story could equally well have been about the application of statistics. I firmly believe that one cannot learn statistics simply by reading a book on the subject. I have thus included a short series of problems which every reader is strongly recommended to complete. The majority of these were set for the Oxford graduate students and used to assess their progress. They claim that the time taken to do these problems was $\sim 40 \pm 10$ hours.

I am grateful to John Lloyd for his valuable comments on a preliminary draft of this book, to many colleagues with whom I have had discussions over the years on statistics, and to all the students whose questions and comments have shown me which parts of the course were in need of further clarification. Finally I want to thank Prof. Yehuda Eisenberg for extending the hospitality of the Physics Department of the Weizmann Institute; the main draft of this book was written in its congenial atmosphere.

L. Lyons, 1985

1

Experimental errors

1.1 Why do we do experiments?

There are basically two different types of results of experiments that scientists perform in order to learn about the physical world. In one type, we set out to determine the numerical value of some physical quantity, while in the second we are testing whether a particular theory is consistent with our data. These two types are referred to as 'parameter determination' and 'hypothesis testing' respectively. (Of course, in real life situations there is a degree of overlap between the two: a parameter determination may well involve the assumption that a specific theory is correct, while a particular theory may predict the value of a parameter.) For example, a parameter determination experiment could consist of measuring the velocity of light, while a hypothesis testing experiment could check whether the velocity of light has suddenly increased by several percent since the beginning of this year.

In this chapter, we are mainly concerned with various aspects of calculating the accuracy of parameter determination type experiments. We will have more to say about hypothesis testing experiments in Chapter 2.

1.2 Why estimate errors?

When we performed parameter determination experiments at school, we considered that the job was over once we obtained a numerical value for the quantity we were trying to measure. At university, and even more so in every-day situations in the laboratory, we are concerned not only with the answer but also with its accuracy. This accuracy is expressed by quoting an experimental error on the quantity of interest. Thus a determination of the velocity of light might yield an answer

$$c = (3.09 \pm 0.15) \times 10^8 \text{ metres/sec.}$$

In Section 1.4, we will say more specifically what we mean by the error

1

of ± 0.15. At this stage it is sufficient to say that the more accurate the experiment the smaller the error; and that the numerical value of the error gives an indication of how far from the true answer this particular experiment may be.

The reason we are so insistent on every parameter determination including an error estimate is as follows. Scientists are rarely interested in parameter determination for its own sake, but more often will use it to test a theory, to compare with other experiments measuring the same quantity, to use this parameter to help predict the result of a different experiment, and so on. Then the numerical value of the error becomes crucial in the interpretation of the result.

For example, let us return to our determination of the velocity of light, and let us use the result to see whether it does indeed look as if its value has increased by, say, a few percent compared with its previously accepted value of 2.998×10^8 metres/sec.†

With a new measurement of 3.09×10^8 metres/sec, do we have evidence for an increase? There are essentially three possibilities.

Possibility 1
If the experimental error is ± 0.15, then the new determination looks perfectly consistent with the old one, and we have no grounds for any new unconventional theory,
i.e. 3.09 ± 0.15 is consistent with 2.998.

Possibility 2
If the experimental error is ± 0.01, then the new determination is inconsistent with the previously accepted value. Hence, on the basis of this new experiment, there does appear to be evidence for the idea that the value of c has increased (or that our experimental result and/or the error estimate are wrong),
i.e. 3.09 ± 0.01 is inconsistent with 2.998.

Possibility 3
With an error of ± 2, the two values of c are consistent but the accuracy of the new experiment is so low that it would be incapable of detecting a change in c of even much more than a few percent,
i.e. 3.09 ± 2 is consistent with 2.998, and with many other values too.

† Since this is an experimental number, it too has an experimental uncertainty, but c has been measured so well that the uncertainty arises in the eighth figure after the decimal point, so we can effectively forget about it here.

Thus for a given value of the parameter, our reaction – 'Conventional physics is in good shape' OR 'We have made a world shattering discovery' OR 'We should find out how to do better experiments' – in this case depends on the numerical estimate of the accuracy of our experiment. Conversely, if we know only that the result of the experiment is that the value of c was determined as 3.09×10^8 metres/sec (but do not know the value of the experimental error), then we are completely unable to judge the significance of this result.

The moral is clear. Whenever you determine a parameter, estimate the error or your experiment is useless.

1.3 Random and systematic errors

There are two fundamentally different sorts of errors associated with any measurement procedure, namely random and systematic errors. The random errors come simply from the inability of any measuring device to give infinitely accurate answers, while the systematic errors are more in the nature of mistakes.

These different types of error are illustrated by considering an experiment involving counters to determine the decay constant λ of a radioactive source. The necessary measurements consist of counting how many decays are observed in a given time in order to determine the decay rate $(-\mathrm{d}n/\mathrm{d}t)$; and of weighing the sample (in order to obtain the number of nuclei n present). Then the decay constant λ is calculated from the formula

$$-\frac{\mathrm{d}n}{\mathrm{d}t} = \lambda n. \tag{1.1}$$

The main random error comes from the fact that there is an inherent statistical error involved in counting random events. Other contributions to the random error come from the timing of the period for which the decays are observed and from the uncertainty in the mass of the sample.

The more obvious possible sources of systematic error are:

(i) The counters used for detecting the decays may not be completely efficient and/or they may not completely surround the source, and so the counting rate may be below the true decay rate.

(ii) The counters may be sensitive to particles coming from other than the source (e.g. cosmic rays), which would give a counting rate above the true decay rate.

(iii) The radioactive source may not be pure (either chemically or

isotopically) and so the number of nuclei capable of decaying may be less than the number as deduced from the mass of the sample.
(iv) There may also be calibration errors in the clock and balance used for measuring the time interval and the sample mass.

A recurring theme in this book is the necessity of providing error estimates on any measured quantity. Because of their nature, random errors will make themselves apparent by producing somewhat different values of the measured parameter in a series of repeated measurements. The estimated accuracy of the parameter can then be obtained from the spread in measured values as described in Section 1.4.2.

An alternative method of estimating the accuracy of the answer exists in cases where the spread of measurements arises because of the limited accuracy of measuring devices. The estimates of the uncertainties of such individual measurements can be combined as explained in Section 1.5 in order to derive the uncertainty of the final calculated parameter. This approach can be used in situations where a repeated set of measurements is not available to employ the method described in the previous paragraph. In cases where both approaches can be used, they should of course yield consistent answers.

The accuracy of our measurements will of course play little part in determining the accuracy of the final parameter in those situations in which the population on which the measurements are made exhibits its own natural spread of values. For example, the heights of ten-year-old children will in general be scattered by an amount which is larger than the uncertainty with which the height of any single child can be measured.

An intermediate situation arises where the observation consists in counting independent random events in a given interval. Although the spread of values will usually be larger than the accuracy of counting (which may well be exact), it is known (see Section 3.2) that for an expected number of observations n, the spread is $\sqrt{n}$.

For systematic errors, however, the 'repeated measurement' approach will not work; if our detector, which we believe counts all particles passing through it, in fact has an efficiency of only 1%, the decay rate will come out too small by a factor of ~ 100 each time we repeat the experiment, and yet everything will look consistent.

Systematic errors can arise on any of the actual measurements that are required in order to calculate the final answer. But they can also appear indirectly in variables which do not explicitly appear in any formulae. Thus, for example, maybe the derivation of the formula being used requires all the apparatus to be at a constant temperature; or we really ought to be performing our experiment in a vacuum instead of in the laboratory atmosphere; or the presence of some impurity is producing spurious results. Ideally, of course, all such effects should be absent. But if it is known that such a distorting effect is present, than at least some attempt can be made to estimate its importance and to correct for it. In effect, we are then converting what was previously a systematic error into what is hopefully only a random one.

In general, there are no simple rules or prescriptions for eliminating systematic errors. To a large extent it is simply a matter of common sense plus experience to know what are the possible dangerous sources of errors of this type. But one possible check that can sometimes be helpful is to use constraints that may be relevant to the particular problem. We give three simple illustrations of this.

Example (a)

We want to know whether a certain protractor has been correctly calibrated. One possible test is to use this protractor to measure the sum of the angles of a triangle. We then repeat this for several triangles and plot a histogram of our results. If the resulting distribution peaks significantly away from 180°, our protractor may be in error.

Example (b)

If a radioactive source emits electrons and photons in coincidence, then (after we have corrected for the inefficiencies of the various counters) we should observe the same number of counts in our electron as in our photon detector. A similar check could be made for a particle whose branching ratios for decays to electrons and to muons are equal.

Example (c)

Nuclear interactions conserve energy and momentum. Thus if we observe a series of interactions and plot the various components of the momentum and the energy imbalance, we expect to obtain distributions which peak around zero. Otherwise there is evidence that something is wrong. For example, the geometrical locations of various pieces of apparatus may not be exactly as we thought, our measurements of the magnitude of the momentum may be systematically in error, there could be some undetected particle taking away significant momentum and energy, etc. (See Fig. 1.1.)

Thus in the above examples, the existence of constraints enables at least

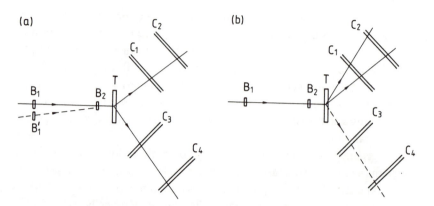

Fig. 1.1. Schematic diagram of apparatus to select examples of a reaction with two outgoing particles. T is the target; B are detectors for the beam particle and C for the outgoing secondaries. (a) Because of alignment problems, counter B_1 is incorrectly thought to be in the position B_1'. This results in the beam tracks' directions being systematically in error. This should become apparent by the failure of p_T, the momentum transverse to the beam, to balance. By moving B_1', its correct position can be found as that which gives the best balance of p_T. An incorrect position for B_1 would given an incorrect value of, for example, the scattering angle θ. (b) In this example, two charged particles passing through C_1 and C_2 are detected, but a neutral particle (shown dashed) passing through C_3 and C_4 is not detected. The momenta of the three outgoing particles balance that of the beam, but with only two of them detected, the observed momenta appear not to balance. This failure to satisfy the momentum constraints enables us to reject this particular event as not being an example of the two body final state that we wish to study (but we clearly need to be careful not to reject events as in (a) above simply because we have made a small error in the assumed positions of the counters).

some check to be performed on the possible existence of systematic effects. In all cases, of course, random effects are also present, and so we do not expect the constraints to be identically satisfied in any particular case. The virtue of using a repeated series of observations is that the spread in the final histogram gives an indication of the magnitude of the random effects,† and hence we can then see whether there is any evidence for a systematic effect too. Random errors are usually more amenable to systematic study, and the rest of this chapter is largely devoted to them.

† The presence of a systematic effect of varying magnitude can broaden an observed distribution as well as shifting it from its expected position. Thus the observed width of an experimental distribution may give an over-estimate of the random effects.

Finally we assert that a good experimental physicist is one who minimises and realistically estimates the random errors of his apparatus, and who reduces the effect of systematic errors to a much smaller level.

1.4 The meaning of σ

In this section we are going to consider in more detail what is meant by the error σ on a measurement. However, since this is related to the concept of the spread of values obtained from a set of repeated measurements, whose distribution will often resemble a Gaussian distribution, we will first have three mathematical digressions into the subjects of (a) distributions in general, (b) the mean and variance of a distribution, and (c) the Gaussian distribution.

1.4.1 Distributions

A distribution $n(x)$ will describe how often a value of the variable x occurs in a defined sample. The variable x could be continuous or discrete, and

Table 1.1. *Examples of distributions*

Character	Limits	x variable	$n(x)$
	$1 \rightarrow \infty$	Integer x	No. of times you have produced a completely debugged computer program after x compilations
Discrete	$1 \rightarrow 7$	Day of week	No. of marriages on day x
	$-13.6\,\mathrm{eV} \rightarrow 0$	Energy of ground and excited states of hydrogen atoms	No. of atoms with electrons in state of energy x in atomic hydrogen at $30\,000°$
	$-\infty \rightarrow \infty$	Measured value of parameter	No. of times measurement x is observed
Continuous	$0 \rightarrow \infty$	Time it takes to solve all problems in this book	No. of readers taking time x
	$0 \rightarrow 24$ hours	Hours sleep each night	No. of people sleeping for time x

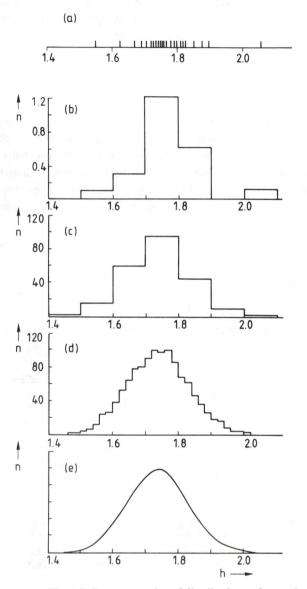

Fig. 1.2. Some examples of distributions of, say, the heights *h* (in metres) of 30-year-old men. (*a*) With only a few observations, each one is represented as a bar at the relevant position along the *h*-axis. (*b*) The data of (*a*) could alternatively be drawn as a histogram, where *n* is the number of men per cm interval of *h*, even though the bin size in *h* is 10 cm. (*c*) A histogram as in (*b*), but with 100 times more observations. (*d*) The same data as in (*c*), but drawn with smaller bins of *h*. *n* is still the number of men per cm interval of *h*. (*e*) For an even larger number of observations and with smaller bin size, the histogram of (*d*) approaches a continuous distribution.

its values could be confined to a finite range (e.g. 0–1) or could extend to $\pm \infty$ (or could occupy a semi-infinite range, e.g. positive values only). Some examples are given in Table 1.1.

Fig. 1.2(a) shows an example of a possible distribution of a continuous variable from a few observations of, say, the height h of 30-year-old men. Since only a few values are available, the data are presented by marking a bar along the h-axis for each measurement. In Fig. 1.2(b), the same data is presented as a histogram, where a fairly wide bin size for h is used and the vertical axis is labelled as n, the number of observations per cm interval of h, despite the fact that the bin size (Δh) used is 10 cm. The actual number of men in a given bin is $n \Delta h$, and the total number of men appearing in the histogram is $\Sigma n \Delta h$. If 100 times more measurements were available, the number of entries in each bin of the histogram would increase by a large factor (Fig. 1.2(c)), but it would now become sensible to draw the histogram with smaller bins, in order to display the shape of the distribution with better resolution. Because we plot $n(h)$ as the number of observations per cm, irrespective of bin size, the overall height of the histogram does not change much when we change the bin size (see Fig. 1.2(d)). Finally, for an even larger number of observations, we could make the bin size so small that the histogram would approximate to a continuous curve (see Fig. 1.2(e)); this could alternatively be viewed as a very good theoretical prediction about the numbers of men of different heights. Again $n(h)$ is to be interpreted in the same way, but now the total number of men appearing in the histogram is $\int n(h) \, dh$.

1.4.2 Mean and variance

In order to provide some sort of description of a distribution such as shown in Fig. 1.2, we need measures of the x-value at which the distribution is centred, and how wide the distribution is. The mean μ and the mean square deviation from the mean σ^2 (also known as the variance) are suitable for this. For a set of N separate measurements such as shown in Fig. 1.2(a), they are defined as†

$$\bar{x} = \Sigma \, x_i/N \tag{1.2}$$

and $\quad s^2 = \Sigma \, (x_i - \mu)^2/N. \tag{1.3}$

† We adopt the convention that the true mean and variance are denoted by μ and σ^2, while the measured mean and variance of a sample are $\bar{x}$ and s^2.

The summations in eqns (1.2) and (1.3) extend over the N values of the sample. In general, the true mean μ is not known, and so eqn (1.3) cannot in fact be used to estimate the variance. Instead it is replaced by

$$s^2 = \frac{1}{N-1} \Sigma \, (x_i - \bar{x})^2. \tag{1.3'}$$

Thus one measurement of a quantity does not allow us to estimate the spread in values, if the 'true' value is not known. (Other problems of using an estimated rather than a true value of a parameter appear in Sections 1.6 and 3.3.)

As an aside, we note that the factor $1/(N-1)$ in eqn (1.3') is required in order to make s^2 an *unbiassed* estimate of the population's variance σ^2, i.e. for a large sample, s^2 will tend to σ^2. If, however, we rewrite the equation as

$$s^2 = \frac{1}{N+k} \Sigma \, (x_i - \bar{x})^2 \tag{1.3''}$$

and set $k = 0$ rather than -1, although the estimate s^2 is now biassed, it turns out that its mean square spread from the correct value σ^2 is smaller† than that for the unbiassed estimate (1.3'). In fact, using $k = +1$ results in that estimate s^2 which has the smallest mean square spread from σ^2. (See problems 5.12 and 6.4.) However, we regard $k = -1$ as preferable. This is because we are usually interested in using the variance in order to judge the significance of the deviation of a measurement $x \pm s$ from some preconceived value y. We then calculate $(x-y)^2/s^2$, which should be distributed like Student's t distribution, but especially if the details of how s^2 was obtained are unavailable, we will be satisfied in assuming that it is Gaussian distributed (see Section 3.3). In that case, it is irrelevant to have a minimum spread estimate for s^2, since we are interested in $1/s^2$, and it is unreasonably small estimates of s^2 which would give trouble. It is because the choices $k = 0$ and $+1$ bias the estimates of s^2 downwards that we regard eqn (1.3') as preferable.

It is most important to realise that s is the measure of how spread out the distribution is, and is not the accuracy to which the mean $\bar{x}$ is determined. This is known to an accuracy better by a factor of $\sqrt{N}$. Thus

---

† The mean square deviation d_0^2 of a set of measurements from the true value t is related to the mean square deviation d^2 from their mean m by

$$d_0^2 = d^2 + b^2,$$

where b is the bias, i.e. $m - t$. As k increases, d^2 decreases but b^2 increases, and there is an optimum value of k which minimises d_0^2.

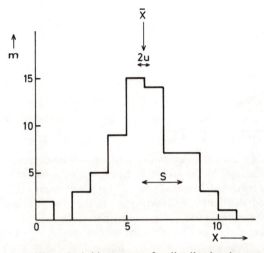

Fig. 1.3. A histogram of a distribution in x. m is the number of entries in each x bin; there are 66 entries in all. The mean $\bar{x}$ and the variance s^2 are estimated as 5.9 and $(2.05)^2$ respectively. The accuracy u to which the mean $\bar{x}$ is determined is smaller than s by a factor of $\sqrt{66}$.

by taking more and more observations of x, the variance s^2 will not change (apart from fluctuations) since the numerator and denominator of eqn (1.3) or (1.3′) grow more or less proportionally; this is sensible since s^2 is supposed to be an estimate of the variance of the overall population, which is clearly independent on the sample size N. On the other hand, the variance of the mean (s^2/N) decreases with increasing N; more data help locate the mean to higher accuracy. (We return to this point, and explain the origin of the $1/N$ factor, in Section 1.5.)

A minor computational point is worth noting. Formula (1.3′) can be written

$$s^2 = \frac{1}{N-1} \Sigma \, (x_i - \bar{x})^2$$

$$= \frac{N}{N-1} [\overline{x^2} - \bar{x}^2], \tag{1.4}$$

where $\overline{x^2}$ is defined in analogy with eqn (1.2) as

$$\overline{x^2} = \Sigma \, x_i^2/N. \tag{1.5}$$

Thus it is not necessary to loop over the data twice (first to calculate $\bar{x}$ and then to obtain s^2 from (1.3′)), but $\overline{x^2}$ and $\bar{x}$ can be calculated in the same loop.

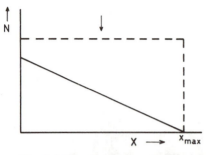

Fig. 1.4. The dashed line corresponds to a uniform distribution in x, with mean value denoted by the arrow. If the experimental efficiency for detecting events falls off linearly from 80% at $x = 0$ to zero at x_{max}, the observed distribution will correspond to the solid line. In order to obtain the correct mean (or any other parameter), the observed distribution must first be corrected for the varying detection efficiency.

Sometimes the measurements are grouped together so that at the value x_j there are m_j events (equivalent to $n_j \Delta h$ in Fig. 1.2(d)). Then simple extensions of eqns (1.2) and (1.3$'$) are

$$\bar{x} = \Sigma \, m_j \, x_j / \Sigma \, m_j \qquad (1.6)$$

and

$$s^2 = \Sigma \, m_j (x_j - \bar{x})^2 / (\Sigma \, m_j - 1). \qquad (1.7)$$

The variance on the mean u^2 is $s^2 / \Sigma \, m_j$ (see Fig. 1.3).

For the continuous distributions of Fig. 1.2(e), these become

$$\bar{x} = \int n(x) \, x \, \mathrm{d}x / N \qquad (1.8)$$

and

$$s^2 = \int n(x) \, (x - \bar{x})^2 \, \mathrm{d}x / N, \qquad (1.9)$$

where

$$N = \int n(x) \, \mathrm{d}x,$$

and where the usual $N - 1$ factor in s^2 has been replaced by N which is assumed to be large for this case.

Another possibility is that the individual measurements have been made with different detection efficiencies, perhaps because they were made with different parts of the apparatus or perhaps because the intrinsic efficiency is a function of the variable x (see Fig. 1.4). Then it is necessary to weight each measurement x_k by a factor w_k which is the reciprocal of the detection efficiency for that particular measurement. Then the mean is given by

$$\bar{x} = \Sigma \, w_k \, x_k / \Sigma \, w_k. \tag{1.10}$$

For the variance and the variance on the mean, we use

$$s^2 = (\Sigma \, w_k (x_k - \bar{x})^2 / \Sigma \, w_k) \times n_{eff} / (n_{eff} - 1) \tag{1.11}$$

and

$$u^2 = s^2 / n_{eff}. \tag{1.12}$$

Here n_{eff} is the *effective* number of events. If the total number of events (after allowing for the detection efficiencies) is $T \pm \delta$, then

$$n_{eff} = \frac{T^2}{\delta^2} = (\Sigma \, w_k)^2 / \Sigma \, w_k^2. \tag{1.13}$$

Thus 100 ± 10 events give us 100 effective events. But with only 4 ± 2 real events with a constant detection efficiency of 4%, the corrected total number is 100 ± 25, and n_{eff} as obtained from eqn (1.13) is 4 (as expected). As a final example, we note that a sample of events of which one has a very low detection efficiency will give a value of ~ 1 for n_{eff}. (See also Comment (iii) in Section 1.6.)

In fact eqns (1.10)–(1.12) can be used for the two previous examples as well.† For the simple case (eqns (1.2) and (1.3′)), the w_k are all unity, while for groups of measurements, the w_k are to be interpreted simply as the number of events with that particular value (x_k).

1.4.3 Gaussian distribution

Since the Gaussian distribution is of such fundamental importance in the treatment of errors, we consider some of its properties now.

The general form of the Gaussian distribution in one variable x is

$$y = \frac{1}{\sqrt{(2\pi)}\,\sigma} \exp\{-(x-\mu)^2/2\sigma^2\}. \tag{1.14}$$

The curve of y as a function of x is symmetric about the value of $x = \mu$, at which point y has its maximum value. (See Fig. 1.5.) The parameter σ characterises the width of the distribution, while the factor $(\sqrt{(2\pi)}\,\sigma)^{-1}$ ensures that

$$\int_{-\infty}^{+\infty} y \, dx = 1. \tag{1.15}$$

The parameter μ is the mean of the distribution, while σ has the following properties.

† With n_{eff} set equal to the total number of events.

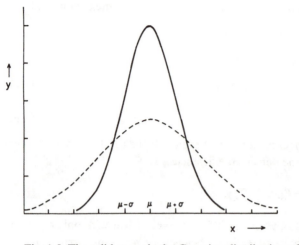

Fig. 1.5. The solid curve is the Gaussian distribution of eqn (1.14). The distribution peaks at the mean μ, and its width is characterised by the parameter σ. The dashed curve is another Gaussian distribution with the same values of μ, but with σ twice as large as the solid curve. Because the normalisation condition (1.15) ensures that the area under the curves is the same, the height of the dashed curve is only half that of the solid curve at their maxima. The scale on the x-axis refers to the solid curve.

(i) The mean square deviation of the distribution from its mean is σ^2 (hence the factor of 2 within the exponent in eqn (1.14)).

(ii) The height of the curve at $x = \mu \pm \sigma$ is $1/\sqrt{e}$ of the maximum value. Since

$$1/\sqrt{e} = 0.607,$$

σ is *roughly* the half width at half height of the distribution.

(iii) The fractional area underneath the curve and with

$$\mu - \sigma \leqslant x \leqslant \mu + \sigma$$

is 0.68.

(iv) The height of the distribution at its maximum is $(\sqrt{(2\pi)}\sigma)^{-1}$. Thus as σ gets smaller, the distribution becomes narrower, and (to maintain the normalisation condition eqn (1.15)) higher at the peak.

By a suitable change of variable to

$$x' = (x - \mu)/\sigma, \tag{1.16}$$

any normal distribution can be transformed into a standardised form

$$y = \frac{1}{\sqrt{(2\pi)}} \exp(-x'^2/2), \tag{1.17}$$

with mean zero and unit variance.

One feature which helps to make the Gaussian distribution of such widespread relevance is the central limit theorem. One statement of this is that if x_i is a set of n independent variables of mean μ and variance σ^2, then for large n

$$y = \frac{1}{n} \Sigma x_i \tag{1.18}$$

tends to a Gaussian distribution† of mean μ and variance σ^2/n. The distribution of the individual x_i is irrelevant. Furthermore, the x_i can even come from different distributions with different means μ_i and variances σ_i^2 in which case y tends to a Gaussian of mean $(1/n) \Sigma \mu_i$ and variance‡ $\Sigma \sigma_i^2/n$. The only important feature is that the variance σ^2 should be finite. If the x_i are already Gaussian distributed, then the distribution of eqn (1.18) is also Gaussian for all values of n from 1 upwards. But even if x_i is, say, uniformly distributed over a finite range, then the sum of a few x_i will already look Gaussian. (This forms the basis of a Monte Carlo technique for generating a Gaussian distribution (see Section 6.4).) Thus whatever the initial distributions, a linear combination of a few variables almost always degenerates into a Gaussian distribution.

The proof of the central limit theorem is given in most standard textbooks (see, for example, Brandt, p. 67). An example of the central limit theorem is given below in Section 1.5.

The Gaussian distribution is discussed again in Section 3.3.

Now that we have concluded our mathematical digressions, we return to our consideration of the treatment of errors.

For a large variety of situations, the result of repeating an experiment many times produces a spread of answers whose distribution is approximately Gaussian; and the approximation is likely to be good especially if the individual errors that contribute to the final answer are small. When this is true, it is meaningless to speak of a 'maximum possible error' of a given experiment since the curve in Fig. 1.5 remains finite for all values of x; the 'maximum possible error' would be infinite, and although this would make it easy to calculate the 'error' on any experiment, it would not distinguish a precision experiment from an inaccurate one.

† For small n, the approximation is better near the peak of the distribution than in the tails.
‡ Provided that the numbers of observations from each of the separate distributions are constant.

It is thus customary to quote σ as the accuracy of a measurement. Since σ is not the maximum possible error, we should not get too upset if our measurement is more than σ away from the expected value. Indeed, we should expect this to happen with about $\frac{1}{3}$ of our experimental results. Since, however, the fractional areas beyond $\pm 2\sigma$ and beyond $\pm 3\sigma$ are only 5% and 0.3% respectively, we should expect such deviations to occur much less frequently.

One word of warning is necessary. Although experimental distributions often approximate to Gaussians near their peaks, in many cases the experimental data will lie above the calculated Gaussian curve when we look out into the tails of the distribution. This can occur because in a whole series of measurements, a few can have lower accuracy than the majority, and hence be more spread out. Even if these lower accuracy measurements themselves lie on a Gaussian distribution with larger σ, the combination of two Gaussian distributions of the same mean but different variances does *not* produce another Gaussian.† This can cause problems with attempts to deduce how unlikely is a deviation of several standard deviations from the expected value. As calculated from a Gaussian distribution fitted to the majority of the data, the probability is likely to be very low. But in reality, it could well correspond to a much more probable fluctuation from a subset of data with larger variance. Thus experiments, whose conclusions are based on interpreting the probability of an unlikely occurrence as being 10^{-8} rather than 10^{-5}, are likely to be unreliable unless their measurement accuracies are very closely controlled (see problem 3.5).

1.5 Combining errors

We are often confronted with a situation where the result of an experiment is given in terms of two (or more) measurements. Then we want to know what is the error on the final answer in terms of the errors on the individual measurements.

1.5.1 Linear situations

As a very simple example, consider

$$a = b - c. \tag{1.19}$$

---

† Note that this is a different type of problem from that discussed in the context of the central limit theorem on p. 15.

To find the error on a, first differentiate

$$\delta a = \delta b - \delta c. \qquad (1.20)$$

If we were talking about maximum possible errors, then we would simply add the magnitudes of δb and δc to get the maximum possible δa. But we have already decided that it is more sensible to consider the root mean square deviations. Then, provided that the errors on b and c are *uncorrelated*,† the rule is that we add the contributions δb and $-\delta c$ in quadrature:

$$\sigma_a{}^2 = \sigma_b{}^2 + \sigma_c{}^2. \qquad (1.21)$$

Two points are worth noting

(i) If in a particular experiment we know that the measurements on b and c were incorrect by specific amounts δb and δc, then the answer would be incorrect by an amount δa, given in terms of δb and δc by the formula (1.20). But the whole point is that in any given measurement we do not know the exact values of δb and δc (or else we would simply correct for them, and get the answer for a exactly), but only know their mean square values σ^2 over a series of measurements. It is for these statistical errors that eqn (1.21) applies.

(ii) For linear combinations like eqn (1.19), it is the errors themselves which occur in eqn (1.21); percentage errors, which are useful for products (see below) are here completely irrelevant. Thus if you wish to measure your height by making independent measurements of the distances of your head and your feet from the centre of the earth, each to 1% accuracy, the final answer will not in general be within 1% of the correct answer; in fact, you may well get a result of -40 miles for your height.

Why do we use quadrature for combining these statistical errors? We look at this in several ways

(a) Mnemonic non-proof

The errors on b and on $-c$ can be in phase with each other to give contributions which add up in δa; or they can be out of phase, so that they partially cancel in δa. So perhaps on average they are orthogonal to each other and hence Pythagoras theorem should be used for obtaining $\sigma_a{}^2$. (See Fig. 1.6.)

We stress that this is not a proof; in particular there is no obvious second dimension in which δb and δc can achieve orthogonality.

† The meaning of 'uncorrelated' becomes clearer later in this section.

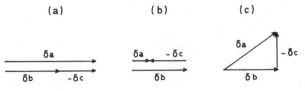

Fig. 1.6. Diagram illustrating the non-proof of formula (1.21). In (*a*) the contributions from δb and from $-\delta c$ are in phase, in (*b*) they are out of phase, while in (*c*) they appear to be in quadrature.

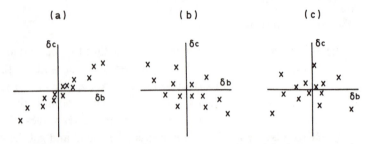

Fig. 1.7. The crosses represent the values of δb and δc for a repeated series of measurements. In (*a*), these errors are strongly correlated, with cov (b, c) being large and positive. The correlation in (*b*) is less pronounced, and cov (b, c) is slightly negative. In (*c*) there is almost no correlation, and the covariance is almost zero.

(b) Formal proof

Given the relationship (1.19), then for the mean square errors†

$$\sigma_a^2 = \langle[a-\bar{a}]^2\rangle$$
$$= \langle[(b-c)-(\bar{b}-\bar{c})]^2\rangle$$
$$= \langle(b-\bar{b})^2\rangle+\langle(c-\bar{c})^2\rangle-2\langle(b-\bar{b})(c-\bar{c})\rangle$$
$$= \sigma_b^2+\sigma_c^2-2\,\text{cov}\,(b, c). \qquad (1.22)$$

The last term involves the covariance of b and c. This is to do with whether their errors are correlated or not. It can be positive, negative or, in the case where the errors are uncorrelated, it will be zero (see Fig. 1.7). Its value is related to the extent to which the exact value of δb in a particular experiment affects that of δc. Some examples of correlations are given below.

If the value of cov (b, c) is zero, then (1.22) reduces to the expected formula (1.21) for uncorrelated variables.

† We use both $\langle x \rangle$ and $\bar{x}$ to represent the average value of x.

(c) The infinitesimal probability argument

We perform an experiment which consists of tossing an unbiassed coin 100 times. We score 0 for each heads and 2 for each tails (i.e. the expectation is 1 ± 1 each time we toss the coin). For the complete experiment, we expect on average to score 100. Now a final score of 0 or 200 is possible, so if we were interested in the maximum possible error, this would be ± 100. But the probability of achieving either of these scores is only $(\frac{1}{2})^{99}$. Thus if we had a team of helpers such that the experiment could be repeated once every second, we would expect to score 0 or 200 once every $\sim 10^{22}$ years. Since the age of the earth is less than 10^{10} years, we can thus reasonably discount the possibility of extreme scores, and thus consider instead what are the *likely* results.

The expected distribution for the final score follows the binomial distribution (see Section 3.1). For 100 throws, this is very like the Gaussian distribution, with mean 100 and $\sigma \sim 10$. We thus have an example of the central limit theorem mentioned in Section 1.4.3; by combining a large number N of variables whose initial distribution consists of two δ-functions, we end up with something very like a Gaussian distribution,† the width of which increases only like $\sqrt{N}$.

(d) Averaging is good for you

We know intuitively that it is better to take the average of several independent measurements of a single quantity than just to make do with a single measurement. This follows from the correct formula (1.21), but not from the incorrect one (1.20).

The average $\bar{q}$ of n measurements q_i each of accuracy σ is given by

$$n\bar{q} = \sum_i q_i. \tag{1.23}$$

Then using (1.21) we deduce that the statistical error δ on the mean is given by

$$n^2\delta^2 = \sum_i \sigma^2 = n\sigma^2,$$

whence

$$\delta = \sigma/\sqrt{n}. \tag{1.24}$$

Thus we have obtained the result quoted in Section 1.4.2 that the error on the mean is known more accurately than the error characterising the distribution by a factor $\sqrt{n}$; this justifies our intuitive feeling that it is useful to average.

† Provided, of course, that we don't look at it with too great a resolution, since this distribution is defined only for integral values, whereas the Gaussian is continuous.

The use of the incorrect formula analogous to (1.20) would have led to the ridiculous result that $\delta = \sigma$.

As can be seen from the formal derivation above, formula (1.21) does not apply when the errors on the variables are correlated. Four simple examples illustrate this. In more complicated situations it is necessary to resort to the use of the error matrix. (See Sections 3.4 and 3.5.)

Example (1)

If
$$a = b + b \tag{1.25}$$

then the two variables on the right-hand side of the equation are completely correlated. Thus if (the first) b is incorrectly measured by an amount δb, then so is (the second) b. So

$$\sigma_a \neq \sqrt{2}\sigma_b$$

but is simply $2\sigma_b$, as can immediately be seen by differentiating the relationship in the form

$$a = 2b. \tag{1.25'}$$

Example (2)

In an experiment, we observe examples of the reactions

$$a + A \rightarrow B + b$$

and
$$a + A \rightarrow B^* + b,$$

where the particle b emerges at some given scattering angle θ. We want to determine the small difference in momenta of particle b in these two reactions, by measuring its curvature in a magnetic field in each case; this will enable us to obtain the excitation energy of the state B^* relative to the ground state B.

The momentum p is determined as

$$p = \frac{0.3Hl^2}{8s},$$

where s is the sagitta of the trajectory (of length l) in the magnetic field H, distances being measured in cm, H in kilogauss and p in MeV/c. Now if the main inaccuracy arises from the measurement error in determining the sagitta, then the momentum error for each track

$$(\Delta p)_s = \frac{8\,\Delta s}{0.3Hl^2}\,p^2$$

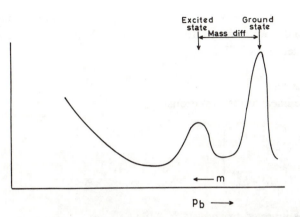

Fig. 1.8. The momentum spectrum of particle b produced in the reaction $a + A \rightarrow B^* + b$, at some fixed scattering angle. This is related to the spectrum of mass m of the state B^*. The peak at the right-hand end of the spectrum corresponds to B^* being produced in its ground state; the other peak corresponds to B^* being in an excited state. If the main contribution to the uncertainty in determining these masses arises from measurement errors, then they are independent and the error in their difference is determined by adding the individual errors in quadrature. If, however, it is the uncertainty in the value of the magnetic field which limits the accuracy, then the mass difference is known much more accurately than is either mass individually.

and these errors are uncorrelated since the measurements error on the two tracks are independent. Hence the error on the momentum difference is given in terms of the individual momentum errors by eqn (1.21).

If, however, the main contribution to the error in momentum arises from an uncertainty in the magnitude of the magnetic field, then the momentum errors are:

$$(\Delta p)_H = \frac{p}{H} \Delta H.$$

Since the possible error in the field value is the *same* for the two tracks, their momentum errors are *correlated*; they are either both too big or both too small. Hence, in the difference of momenta, this uncertainty to a large extent cancels. Thus

$$[\Delta(p_1 - p_2)]_H = (p_1 - p_2)\Delta H / H,$$

which is smaller than the individual uncertainties Δp_H since $p_1 - p_2$ is small compared with either of the ps (and of course also smaller than would have been obtained from the individual Δp_H, assuming that the errors are uncorrelated).

More realistically, we would observe the distribution in momentum of b in order to deduce the excitation spectrum of B* (see Fig. 1.8), but the above considerations still apply.

Example (3)

Consider the linear transformations

$$\left.\begin{aligned} x &= \frac{1}{\sqrt 2}(a+b), \\ y &= \frac{1}{\sqrt 2}(a-b), \end{aligned}\right\} \tag{1.26}$$

which can be inverted to give

$$\left.\begin{aligned} a &= \frac{1}{\sqrt 2}(x+y), \\ b &= \frac{1}{\sqrt 2}(x-y). \end{aligned}\right\} \tag{1.27}$$

Now if the statistical errors on a and b are such that

$$\sigma_a > \sigma_b, \tag{1.28}$$

then

$$\sigma_x = \sigma_y = \sqrt{[(\sigma_a{}^2 + \sigma_b{}^2)/2]}, \tag{1.29}$$

but they are not independent of each other. If we (incorrectly) assume that σ_x and σ_y are independent of each other, then we would deduce from (1.27) that σ_a and σ_b were both equal to $\sqrt{(\sigma_a{}^2 + \sigma_b{}^2)}$, which is clearly incorrect. (This case is considered again in terms of error matrices as example (iii) in Section 3.6.)

Example (4)

Suppose that the high energy reaction

$$K^-p \to \bar{K}^{\circ}p\pi^- \tag{1.30}$$

is in part directly produced, and is partly produced via resonant channels corresponding to intermediate states $\bar{K}^*(890)\,p$, $\bar{K}^*(1420)\,p$, and $Y^*(1760)\,\pi$. The fractions of the final state produced according to these channels are denoted by $f_0 \ldots f_3$ respectively. If we use the experimental data to determine these fractions then their errors will be correlated because of the constraint

$$\Sigma f_i = 1. \tag{1.31}$$

In particular, if we calculate Σf_i from the measured quantities, then we should get the answer of unity; the correlations between the f_i ensure that the overall statistical error on Σf_i is zero.

As a final comment on correlations, we note that the following situation is possible

$$\left.\begin{array}{l} a \text{ is correlated with } b, \\ b \text{ is correlated with } c, \\ a \text{ is } not \text{ correlated with } c. \end{array}\right\} \tag{1.32}$$

Thus 'correlation' is by no means to be thought of as an equals sign. A simple example illustrating the relations (1.32) is achieved by assuming that a and c are each uniformly distributed over the range 0–1 but are uncorrelated, and that

$$b = a + c.$$

This achieves the correlation of b with either a or c. An appreciation of which kinematic variables are correlated with each other and which are independent† is of great importance in assessing the comparison of physical models with experimental data. We return to this point in Section 6.6.2.

1.5.2 Non-linear situations

So far we have been considering how to combine errors when the formula relating the answer to the measurements is linear. In other cases, the correct answer can be achieved by first differentiating, then collecting together the terms of each independent variable and finally adding these in quadrature, i.e. for $y(x_1, x_2, \ldots, x_n)$,

$$\sigma_y^2 = \sum_{i=1}^{n} \left(\frac{\partial y}{\partial x_i}\right)^2 \sigma_{x_i}^2 \tag{1.33}$$

Thus, for example, if

$$a = b^r c^s, \tag{1.34}$$

where r and s are known constants, then assuming the errors on b and c are uncorrelated

$$\left(\frac{\sigma_a}{a}\right)^2 = r^2 \left(\frac{\sigma_b}{b}\right)^2 + s^2 \left(\frac{\sigma_c}{c}\right)^2, \tag{1.35}$$

i.e. the fractional errors on b and c are combined to give the fractional error on a.

† The reader is invited to think of a set of kinematic variables which possess the property (1.32).

As before, when dealing with ratios or products we must be careful about correlations, whose effect is to invalidate eqn (1.35). Thus, for example, consider a measurement of the cross-sections† σ_i for two different processes in a given experiment (e.g. for interactions producing, respectively, two or four charged particles), and also their ratio q. For thin targets, the cross-sections σ_i are determined as

$$\sigma_i = \frac{n_i}{tB}$$

where n_i are the numbers of observed interactions of the types of interest, B is the beam intensity and t is the thickness of the target measured as the number of nuclei per unit area of the target. If the major contribution to the error in σ arises from the statistical uncertainty in the number of observed interactions n_i (see Section 3.2), then the errors $\delta\sigma_1$ and $\delta\sigma_2$ are independent, and the error on the ratio is given by eqn (1.35) as

$$\left(\frac{\delta q}{q}\right)^2 = \left(\frac{\delta\sigma_1}{\sigma_1}\right)^2 + \left(\frac{\delta\sigma_2}{\sigma_2}\right)^2. \tag{1.36}$$

But if the major uncertainty in σ_i is due to a poorly determined beam flux, then although neither cross-section is well determined absolutely, this uncertainty cancels in the ratio q, i.e. the errors on σ_1 and σ_2 are correlated such that δq is very much smaller than would be given by eqn (1.36).

When correlations are present between b and c in eqn (1.34), eqn (1.35) becomes

$$\left(\frac{\sigma_a}{a}\right)^2 = r^2 \left(\frac{\sigma_b}{b}\right)^2 + s^2 \left(\frac{\sigma_c}{c}\right)^2 + 2rs\,\frac{\mathrm{cov}\,(b,c)}{bc}. \tag{1.37}$$

Again, more complex situations are probably best dealt with by the error matrix techniques described in Section 3.5.

For the linear case, the error formula (1.21) applies whatever the magnitudes of the individual errors, but for non-linear cases this is not true and formulae like eqn (1.35) are applicable only when the individual errors are small. As a counter-example, consider determining a ratio (perhaps of two decay modes of a given resonant state) when the denominator has a large fractional error. Thus, for example,

$$a = b/c$$

$$= \frac{100 \pm 10}{1 \pm 1}$$

$$= 100 \pm ?$$

† The reader should be aware that σ is used as the symbol for both the statistical variance and the physical cross-section.

The error as calculated from the formula (1.35) gives an answer of ± 100. Is this realistic, or could you think of a way of quoting the error on a which gives a better idea of our knowledge of the ratio?

1.6 Combining results of different experiments

When several experiments measure the same physical quantity and give a set of answers a_i with different errors σ_i, then the best estimates of a and its accuracy σ are given by

$$a = \Sigma (a_i/\sigma_i^2)/\Sigma (1/\sigma_i^2) \tag{1.38}$$

and

$$1/\sigma^2 = \Sigma (1/\sigma_i^2). \tag{1.39}$$

Thus each experiment is to be weighted by a factor $1/\sigma_i^2$. In some sense, $1/\sigma_i^2$ gives a measure of the information content of that particular experiment. The proof of formulae (1.38) and (1.39) can be found in any standard textbook. (See, for example Orear, p. 8; Brandt, p. 97; or problem 5.7(a).)

Some comments on formulae (1.38) and (1.39) are in order.

Comment (i)

The formulae agree with common sense when the different experiments have achieved different accuracies σ_i by using the same apparatus but by repeating the measurements n_i times and averaging, in which case the σ_i are proportional to $1/\sqrt{n_i}$ (see Example (d) in Section 1.5).

Then the formulae (1.38) and (1.39) become

$$a = \Sigma n_i a_i/\Sigma n_i \tag{1.40}$$

and

$$N = \Sigma n_i. \tag{1.41}$$

The first of these says that each of the original measurements of equal accuracy is to be weighted equally; the second is self-evident. (See problem 1.4.)

Comment (ii)

We now analyse a situation where the use of the averaging formulae appears to be incorrect.

We assume we are trying to measure a counting rate in a particular experiment where the rate stays essentially constant. (For example, we might be counting cosmic rays passing through our apparatus, decays from a radioactive source of long lifetime, or interactions produced by a beam in our target.) Let us assume that we observe 1 ± 1 counts† in the first hour and 100 ± 10 counts in the second hour. (N.B. Of course, as we have emphasised before, no self-respecting experimental scientist should simply shut off his mind and blindly apply statistical formulae without thinking. In this particular case, it would be criminal to average the two measurements which are so clearly inconsistent with each other; it would be essential to find out what has gone wrong. Our excuse for proceeding is that we have used very different rates simply to produce a more dramatic numerical effect.)

The formula for combining experiments gives the best estimate of the rate as

$$2 \pm 1 \text{ counts per hour,} \tag{1.42}$$

since the first hour is weighted by a factor of 100 with respect to the second. But common sense gives the average as

$$50.5 \pm 5 \text{ counts per hour.} \tag{1.43}$$

Not only is this obvious, but it is also correct as we now show.

Where we have made the mistake is that in using the averaging formulae (1.38) and (1.39), we are required to use the *true* values σ_i for each hour, whereas in fact we have used instead our estimates s_i. Since we assume that the particle flux and the apparatus do not vary over the two hours, the *true* counting rates are the same (that is why we want to combine them to get an improved average value), and so are the *true* variances σ_i^2. Hence we should give the individual measurements equal weight, which then results in the standard formulae (1.38) and (1.39) giving the correct result (1.43).

One moral from this is that it is important to be careful in combining experiments on, for example, rare decay modes of nuclei or particles. These are likely to be dominated by statistical errors associated with the limited number of observed decays. Hence it is essential to impose as a constraint that the expected decay branching ratio is the same for all experiments. This then enables us to assign sensible weights to each experimental determination, rather than to give unduly large weights to experiments which observe lower values of the decay rate and hence have smaller estimated errors.

† The reason why the error is $\sqrt{n}$ in this type of situation is explained in our discussion of the Poisson distribution in Section 3.2.

Comment (iii)

A source emitting particles is completely surrounded by two hemispherical counters, one of 100% efficiency and the other of 4%. The observed numbers of counts in these detectors in unit time are 100 ± 10 and 4 ± 2; after correcting for the counter inefficiency, the latter number becomes 100 ± 50. We wish to calculate the best estimate of the total decay rate.

If we know that each counter subtends a solid angle of 2π as seen from the source, then the optimum procedure is to find the average rate according to formulae (1.38) and (1.39) as 100 ± 9.8, which then gives a total decay rate of

$$200\pm19.6. \tag{1.44}$$

This answer also seems plausible, as from formula (1.13) it corresponds to 104 effective events, consisting of the 100 observed events from the first counter and the 4 from the second.

But if we do not know the relative solid angles of the two counters, we have to estimate them from the ratio of the number of counts in the two counters (after correcting for counter inefficiencies). Then the best estimate of the total number of counts is simply the sum of the two numbers, i.e.

$$200\pm51. \tag{1.45}$$

This result is of considerably lower accuracy than the previous estimate, eqn (1.44). Thus we see that extra information (the knowledge of the relative solid angles in the first case) results in improved accuracy of the answer.

For another example of this approach, see problem 1.5.

Problems

1.1 (a) A set of 13 measurements are made on a physical quantity. The following values are obtained:

$0, 1, 2, 3, ..., 12.$

Estimate the mean $\bar{x}$, the spread σ and the accuracy of the mean u.

(b) A new set of 36 measurements are made with the result that the values

$0, 1, 2, ..., 5, 6, 7, ..., 11, 12$

occur $0, 1, 2, ..., 5, 6, 5, ...,\ 1,\ 0$ times respectively. Estimate $\bar{x}, \sigma$ and u.

(c) The function $y(x)$ is defined as

$$y = 1/L \quad \text{for} \quad 0 \leqslant x \leqslant L$$

$$= 0 \text{ otherwise.}$$

Find the average value of x, and the spread σ for this distribution.

(d) Repeat the problem of (c) above, but for the function

$$y = 4x/L^2 \quad \text{for} \quad 0 \leqslant x \leqslant L/2$$

$$= 4(L-x)/L^2 \quad \text{for} \quad L/2 \leqslant x \leqslant L$$

$$= 0 \text{ otherwise.}$$

(e) Compare the answers for (a) and (c), and for (b) and (d).

1.2   A man lives in a rectangular room for which he wants to buy carpet and wallpaper. The required quantities of these will be proportional to the floor's area and perimeter respectively. He thus measures the floor, and finds that its dimensions are $l \pm \sigma_l$ and $b \pm \sigma_b$, with the errors being uncorrelated. Find the errors on the area and on the perimeter, and show that they are correlated.

1.3   By measuring yourself with three different rulers, you obtain the following estimates of your height

$$165.0 \pm 1.0, \quad 164.9 \pm 0.5 \quad \text{and} \quad 165.6 \pm 0.1 \text{ cm.}$$

What is the best estimate of your height, and how accurate is it?

1.4   Three schoolboys A, B and C perform a pendulum experiment with the same apparatus in order to determine the acceleration due to gravity g. An individual measurement consists of timing 100 swings of the pendulum, and this is what A does. However, B does this twice and averages the two values to obtain an improved answer, while C takes the average of ten sets of swings. If A's answer has an uncertainty σ_a, what are the expected accuracies of B's and of C's determinations? (Assume that the dominant error is the random one associated with timing the swings.)

The schoolmaster now takes the three students' determinations ($a \pm \sigma_a$, $b \pm \sigma_b$ and $c \pm \sigma_c$) and uses the prescription (1.38) and (1.39) to obtain his estimate of g and its error. Show that these are identical with what the teacher would have obtained by taking all 13 individual measurements and averaging them, without regard to which student had performed which determination.

1.5 In high energy physics experiments, K°s are often produced in the liquid of a bubble chamber and may then be seen to decay within the chamber, but because of their high energy, they can decay beyond it and hence be unobserved. In order to correct for this, it is conventional to weight each observed decay by a factor $(1 - \exp(-ml/cp\tau))^{-1}$ to allow for the efficiency of observing decays of K°s of momentum p when they are produced at a distance l away from the end of the chamber's visible volume. (m and τ are the mass and lifetime of the K°.) For K°s produced near the end of the chamber, l is small and the weights can become very large. It has thus been customary to throw away events with K°s produced too near the end of the chamber, in order to avoid diluting the quality of good events (with weights ~ 1) by these poorer data. This clearly cannot be the optimal way for analysing the data.

Devise a better procedure, making use of the fact that the production positions of the K°s are uniformly distributed throughout the length L of the bubble chamber. (You may simplify the problem by assuming that all the K°s are produced in the forward direction so that the potential length l for a K° decay is given by

$$l = L - x,$$

where x is the production position, measured from the beginning of the chamber – see the diagram.)

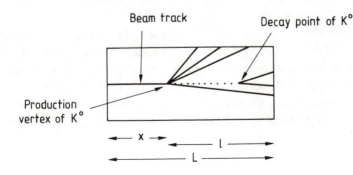

2

Probability and statistics

2.1 Probability

What do we mean by 'probability'? In many situations we deal with
experiments in which the essential circumstances are kept constant, and
yet repetitions of the experiment produce different results. Thus the result
of an individual measurement or trial may be unpredictable, and yet the
possible results of a series of such measurements have a well-defined
distribution. The probability p of obtaining a certain specified result on
performing one of these measurements is then simplest to visualise as the
ratio

$$p = \frac{\text{number of occasions on which that result occurs}}{\text{total number of measurements}}. \qquad (2.1)$$

Example (a)
We throw a die and want to score four. Because the die has six faces and
because of the symmetry of the situation (assuming the die is unbiassed)
we expect to be successful in $\frac{1}{6}$ of our attempts. In this case our experiment
results in a discrete distribution with only six different possibilities, each
equally probable.

Example (b)
A radioactive source situated in a magnetic field decays isotropically in
space. A counter which detects one of the decay products subtends a solid
angle of $d\Omega$ steradians as seen by the source, and can be set at an angle
θ with respect to the direction of the magnetic field. The probability that,
in any given decay, the decay product will pass through the detector is
$d\Omega/4\pi$, independent of θ. Here the distribution in θ is continuous, and
constant.

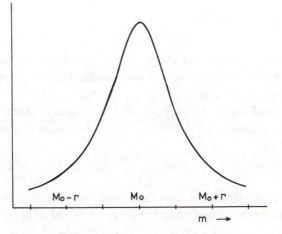

Fig. 2.1. The Breit–Wigner distribution $B(m)$, as an example of a probability function of a continuous variable. The relative probability of obtaining a mass within a small range of any specific value m is proportional to the height of the curve at that particular point. Thus we are most likely to observe values of m near the central value M_0; masses that differ from M_0 by much more than Γ are unlikely.

Example (c)

We count the number n of throws of a die required until we score six. Clearly, lower values of n are more likely than higher values. This time we have a discrete distribution with an infinite number of possibilities (n can take any integral value from one upwards), and the relative probabilities are different.

Example (d)

An excited state is produced in a nuclear reaction, but it has a very short lifetime. Thus its mass m is not uniquely defined, but is characterised by a Breit–Wigner distribution $B(m, M_0, \Gamma)$ whose parameters are the mass M_0 and width Γ of the resonant state (see Fig. 2.1). We may be interested in the fraction δf of interactions with masses in the range m to $m + \delta m$, i.e. $\delta f = B(m, M_0, \Gamma)\,\delta m / \int B\,dm$. Then $\delta f/\delta m$ is known as the probability function and gives the relative probabilities of observing different values of the relevant variable, in this case the mass. Here we are dealing with a continuous distribution, where different values of the parameter have different probabilities.

2.2 Rules of probability

Rule 1

The probability of any particular event occurring is a number between zero and one inclusive. A probability of zero implies that that particular event never happens, while a value of one implies that it always occurs (see formula (2.1)). Thus for our die, the probability of throwing a seven is zero, of obtaining any number less than 10 is unity, and of obtaining an even number is $\frac{1}{2}$.

Rule 2

The probability $P(A+B)$ that at least one of the events A or B occurs is given in terms of the individual probabilities of $P(A)$ and $P(B)$ by

$$P(A+B) \leqslant P(A)+P(B). \tag{2.2}$$

The equality applies if the events A and B are exclusive, i.e. if the occurrence of A precludes that of B. Thus the probability of obtaining a three or an even number when we throw a die is

$$P(3 \text{ or even}) = P(3)+P(\text{even})$$

$$= \tfrac{2}{3},$$

whereas the probability of obtaining either a number below 3.5 or an even number is $\frac{5}{6}$, i.e.

$$P(\text{below } 3.5 \text{ or even}) < P(\text{below } 3.5)+P(\text{even})$$

since

$$\tfrac{5}{6} < \tfrac{1}{2}+\tfrac{1}{2}.$$

We see that when A and B have common elements, the inequality applies.

Rule 3

The probability $P(AB)$ of obtaining both A and B is given by

$$P(AB) = P(A/B)\,P(B)$$

$$= P(B/A)\,P(A), \tag{2.3}$$

where $P(A/B)$ is the probability of obtaining the event A given that B has occurred. It is known as 'the conditional probability of A, given B'.

Thus we obtain the very sensible result that

$$P(A/B) = \frac{P(AB)}{P(B)}, \tag{2.4}$$

i.e. the conditional probability of A is obtained by dividing the number of times that both A and B are observed together by the total number of times that B occurs.

If the occurrence of B does not affect whether or not A occurs, then

$$P(A/B) = P(A) \tag{2.5}$$

and A and B are said to be independent. From eqns (2.3) and (2.5), we then obtain that

$$P(AB) = P(A) P(B) \tag{2.6}$$

for independent events.

We illustrate this by two examples of independent events.

Example (a)
A = It is Sunday.

B = It is raining.

The probability of its being rainy on a Sunday is the same as that for any other specific day of the week, and so A and B are independent. (But if we replaced 'Sunday' by 'December' in our definition of A, then we would lose the property of independence, since most places have a higher than average rainfall at that time of the year.)

Example (b)
A and B correspond to two different beam tracks interacting in a target. Whether the first track interacts or not does not affect the other. This is true even if the beam particles are of different types with different interaction probabilities.

We now have some examples of events that are not independent.

Example (c)
If each week you spend 42 hours in your laboratory, the probability at any moment that your head is in the laboratory is 0.25. Similarly, the probability of your feet being in the laboratory is also 0.25. But the probability of both your head and your feet being there is again 0.25, and is not $\frac{1}{16}$, as would be obtained from eqn (2.6). Here the non-independence rises from the correlation between your head and your feet: they are constrained to be less than about 1.5 metres apart.

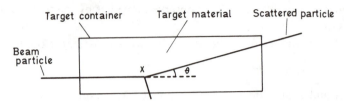

Fig. 2.2. A charged particle enters a target and interacts at x, being scattered through an angle θ. Because of the energy loss in passing through the target, interactions at the right-hand end of the target occur at a lower beam energy than those at the left, and hence the scattering angular distributions may be slightly different.

Example (d)

Abram and Lot were standing at a road junction. The probability of Lot taking the left-hand road was 0.5. The probability of Abram taking the left-hand road was also 0.5. But the probability of both of them going to the left was zero. (The reader interested in further details should consult Genesis **13**.9.)

Example (e)

The Fermi theory allows us to calculate the electron spectrum in β decay, e.g.

$$n \rightarrow p + e^- + \bar{\nu}.$$

So we can obtain the probability of the electron having a certain high fraction (say, $\geqslant \frac{3}{4}$) of the available energy. We can also do a similar calculation for the $\bar{\nu}$. But the probability of their *both* having high energies is zero, since they are constrained by energy conservation to share the total available energy.

Finally we have an example of almost independent variables.

Example (f)

Consider a beam particle interacting in a target (see Fig. 2.2). Then

> A = position of interaction in target.

> B = scattering angle in that interaction.

If the beam momentum is not too high, then the energy loss as the beam passed through the target may be significant. Now, in general, scattering angular distributions depend on energy, and so will change slightly across the target. Thus the scattering angle and the interaction position are not completely independent, although at high energies or for short targets this lack of independence may in practice be unimportant.

The question of whether or not physical quantities are independent is similar to the subject of correlated errors, which was discussed in Section 1.5.

2.3 Statistics

In probability theory, we in general start with some well defined problem, and calculate from this the possible outcomes of a specific experiment. We thus proceed from the theory to the data. In 'statistics' we try to solve the inverse problem of using the data to enable us to deduce what are the rules or laws relevant for our experiment. Some examples illustrating this relationship are given in Table 2.1.

Since this book deals mainly with the actual problem of analysing experimental data, we are concerned more with statistics. It should already be clear, however, that statistics depends on the results of probability theory, and so that is why some considerations of probability are to be found here.

The basic problem an experimentalist faces is how to summarise his data efficiently. Usually this will consist either of checking whether the data are consistent with a given hypothesis ('hypothesis testing') or in using the data to determine a parameter of a model ('parameter estimation').

(a) Parameter estimation

Here we are required to determine the value of some parameter *and its error*. Without an estimate of the error, the experiment is meaningless. This was already discussed in detail in Section 1.2.

The estimation of an error Δp is usually more difficult than that of the corresponding parameter p, but it is just as important and *must* be done.

The parameter and its error should also be determined in an *unbiassed* and *efficient* manner.

Unbiassed means that the method we are using on average will give us the correct result. For example, we might try to measure the Maxwellian distribution of velocities in a gas by observing molecules coming out of a small hole in a hot oven. But faster moving molecules have a greater probability of getting out of the oven than do slower ones, and so the emerging beam is not typical of the velocity distribution inside the oven,

Table 2.1. *The relation between probability and statistics*

Probability	Statistics (P.D. = param.determ.; H.T. = hypothesis testing)
From theory to data	From data to theory
e.g. (a) Toss a coin. If $P(\text{heads}) = \frac{1}{2}$, how many heads in 50 tosses?	Observe 27 heads in 50 tosses. What is value of $P(\text{heads})$? [P.D.]
(b) Toss 50 coins each 100 times. If $P(\text{heads})$ identical for all coins, what relationships between numbers of heads for different coins?	From observed numbers of heads, are data consistent with same $P(\text{heads})$ for all coins? [H.T.]
(c) Given a sample of μ mesons of known polarisation, what is forward/backward decay asymmetry?	Of 1000 μ mesons, 600 observed to decay forwards. What is μ polarisation? [P.D.]
(d) If parity is conserved, what restrictions on decay distributions?	Given a particular decay distribution, is parity conserved? [H.T.]
In all the above cases, we can calculate either some parameters of the distribution (e.g. mean, spread), but it is better to calculate the complete distribution.	In hypothesis testing, we can compare value of parameter with theory, or better compare the whole distribution.
	Here we use the observed data (i) to deduce some quantity of interest *and its error*; or (ii) to check a theory at some *confidence level*.

i.e. the observed distribution is biassed (and so we must compensate for this effect if our results are to be meaningful). In this example, the potential bias arises from the design of the experiment, but similar effects can be the result of the statistical methods used.

The word 'efficiency' has a technical meaning, and refers to the fact that you should derive as much information as possible from your data. Thus if you are trying to determine the mass of a resonant state (see Fig. 2.1), it is not a good idea to take the average of the highest and lowest observed values. We should rather look for a method which gives us as small as possible an error on our parameter (in this case the mass of the resonance).

On the other hand, it is also important to consider the personal aspects of 'efficiency'. Statistics is, as far as we are concerned, the servant of the sciences, and we are interested in obtaining answers to our problems as quickly as possible. Thus we do not want to spend a very long time† using a very sophisticated method of solving a problem when a simpler method is almost as good.

(b) Hypothesis testing

Here, for example, we are trying to answer a question like: 'Is the shape of the electron's energy spectrum as observed in a particular β-decay process in agreement with the Fermi theory?' The answer will not be a simple yes or no, but will also contain a statement of how confident we are of our answer. A statement that at the 95% confidence level the data are inconsistent with the Fermi theory should mean that our acceptance criteria for the hypothesis are such that, if we performed 100 similar experiments on nuclei whose β-decay was determined by the Fermi theory, only five or so would appear to be inconsistent. Just as with the determination of errors in 'parameter estimation', so here it is essential to make such a confidence statement.

In fact parameter determination and hypothesis testing are not really as distinct as suggested above. Thus in order to determine a parameter, we have usually already assumed that some hypothesis is necessarily true; while a specific hypothesis may involve a parameter whose value needs to be determined as part of the hypothesis testing procedure. For example, the distribution of β-decay particles with respect to a magnetic field may be predicted to be of the form $1 + b \cos \theta$. Hypothesis testing then checks whether the data are consistent with such a form, which can only be performed after we have obtained a value for b; while parameter determination provides a value for b, which will of course be meaningless if the form of the distribution is incorrect.

For both parameter estimation and hypothesis testing, 'statistics' provides a set of rules which enable us to calculate various numbers of relevance. But it is not simply a case of blindly following a set of cookery instructions – our own personal judgement is involved too. Thus in fitting a Breit–Wigner formula to an experimental mass distribution, it is up to us to decide over what mass range the fit should be performed, whether

† This comment applies both to the amount of our own time spent in thinking, and to the amount of computer time that almost inevitably is involved in solving any but the simplest of problems. Again a balance must be found between these two times – often a little more thought can save a lot of computer time and make you more popular with your colleagues.

there is some non-resonant background contribution and what functional form should be used for it, whether all the data are of good resolution or whether some should be discarded, etc. Thus, whereas problems in statistics books may have unique answers, real life situations require us to make subjective choices.

Problems

2.1   A card is withdrawn at random (a) from a standard pack of 52 playing cards or (b) from another standard pack to which a joker has been added, i.e. 53 cards in all. For each case (a) and (b), calculate

(i) the probability P_2 that the selected card is a 2,
(ii) the probability P_s that the selected card is a spade,
(iii) the probability P_{2s} that the selected card is the 2 of spades, and compare the value of $P_2 \times P_s$ with P_{2s}. Explain the comparisons.

2.2   A random person, selected from a bus queue, was wearing old jeans, a torn sweater and muddy shoes. A statistics student was asked to guess the probability that the man was a bank clerk; his answer was 0.001. The question was then changed to assessing the probability that he was a bank clerk who played the guitar in a pub at week-ends; this time his estimate was 0.005. Comment.

3

Distributions

We have already seen that in many situations, the result of repeating an experiment many times does not lead to the same result, but produces a distribution of answers. The form of the distribution depends on the nature of the experiment. In this chapter, we discuss only the binomial, Poisson and Gaussian distributions. Other distributions are discussed, for example in Chapter 4 of Eadie, or Chapter 5 of Brandt.

This chapter thus deals with probability theory, but the results are of importance for statistics.

3.1 Binomial

The binomial distribution applies to situations where we conduct a fixed number N of independent trials, each of which has only two possible outcomes – success (with probability p) or failure (with probability $1-p$). Then the probability of obtaining r successes is given by

$$P(r) = \frac{N!}{r!(N-r)!} p^r (1-p)^{N-r} \tag{3.1}$$

for values of r from 0 to N inclusive. The derivation of this can be found in any standard text on probability theory, but essentially the form of (3.1) is obvious. The p^r term is the probability of obtaining successes on r specific attempts, and the $(1-p)^{N-r}$ factor is that for failures on the remaining $N-r$ trials. But this corresponds to only one ordering of the successes and failures; the factorial term then gives the number of permutations of r successes and $N-r$ failures.

Some examples of when we would expect the binomial distribution to apply are:

Situation (i)
We throw a die N times. What is the probability of obtaining the 6 face upwards on exactly r occasions?†

---

† The related problem of the probability of requiring exactly M throws of the die in order to obtain a fixed number r of 6's gives rise to the inverse binomial distribution.

Situation (ii)

A histogram, whose expected distribution is known (for example, a mass distribution which follows a Breit–Wigner shape of given central mass and width), contains exactly N events. What is the probability of having exactly r events in a specific bin of the histogram?†

Situation (iii)

The angles that the decay products from a given source make with some fixed axis are measured. If the expected distribution is known (for example, the source may decay isotropically), what is the probability of observing r decays in the forward hemisphere (i.e. with $\theta < \pi/2$ from a total sample of N decays?

For the binomial distribution (3.1), the expectation value of the number of successes r is

$$\bar{r} = \Sigma \, rP(r) = Np. \tag{3.2}$$

This is not too surprising since we have N independent trials, each with a probability p of success. We can thus estimate p from an observed distribution as the fraction $\bar{r}/N$.

The variance of the distribution is given by

$$\sigma^2 = Np(1-p). \tag{3.3}$$

When p is unknown, we can use the estimate for it to estimate σ^2 by

$$s^2 = \frac{N}{N-1} \, N \, \frac{\bar{r}}{N}\left(1 - \frac{\bar{r}}{N}\right). \tag{3.4}$$

(Compare the $1/(N-1)$ factor in formula (1.3′).) The derivation of formulae (3.2) and (3.3) is simple but tedious.

We see from (3.2) and (3.3) that

$$\sigma^2 \leqslant \bar{r}, \tag{3.5}$$

where the equality holds only for $p = 0$. Thus in general the variance is smaller than the mean.‡ This is because there is an upper limit (of N) imposed on r, which thus reduces the spread of the r-distribution. For example, if p is equal to one, then the only non-zero value of P is when $r = N$

---

† The probability distribution for obtaining r_1 events in the first bin, r_2 in the second, etc., is given by the multinomial distribution, which is the extension of the binomial distribution to the case when the outcome of each trial can have more than two types of results. The binomial distribution is a special case of this, and applies to situations where each trial can lead to one of two results, usually called 'success' and 'failure'.

‡ Contrast the Poisson distribution, where these quantities are equal (see Section 3.2).

(we only have successes) and hence σ is zero. If p is very small, the upper limit N is unimportant and $\sigma^2 \sim \bar{r}$. Similarly for p almost equal to unity, the constraint on the number of *failures* is unimportant and so the variance on the number of failures (which of course is equal to the variance on the numbers of success†) is approximately equal to the number of failures.

Some examples of the binomial distribution are shown in Fig. 3.1, (*a*) for fixed N, (*b*) for constant p, and (*c*) for Np constant.

The binomial distribution itself is not of great application to nuclear physics, since we rarely work with a fixed number of events. Its main use is in the limits

(i) $p \to 0$, $N \to \infty$ but $Np = \lambda$ (constant), when binomial $\to$ Poisson (see Section 3.2, and Figs. 3.1(*c*) and 3.2).

(ii) $N \to \infty$, $p =$ constant, when binomial $\to$ Gaussian (see Section 3.3, and Fig. 3.1(*b*)).

These relationships are illustrated later in Fig. 3.5.

3.2 Poisson

This is the limit of the binomial distribution as

$$
\left.
\begin{aligned}
& N \to \infty, \\
& p \to 0, \\
& Np = \text{constant} = \mu t = \lambda.
\end{aligned}
\right\}
\tag{3.6}
$$

Then

$$
P_r(t) = \frac{(\mu t)^r}{r!}\, e^{-\mu t}
\tag{3.7}
$$

$$
= \frac{\lambda^r}{r!}\, e^{-\lambda}.
\tag{3.7'}
$$

This is the probability of observing r independent events in a time interval t, when the counting rate is μ and the expected number of events in the time interval is λ.

The approach of the binomial to the Poisson distribution as N increases is shown in Fig. 3.2.

The mean value of r for a variable with a Poisson distribution is λ and so is the variance. This is the basis of the well known

$$
n \pm \sqrt{n}
\tag{3.8}
$$

† This is why the variance is symmetric between p and $(1-p)$.

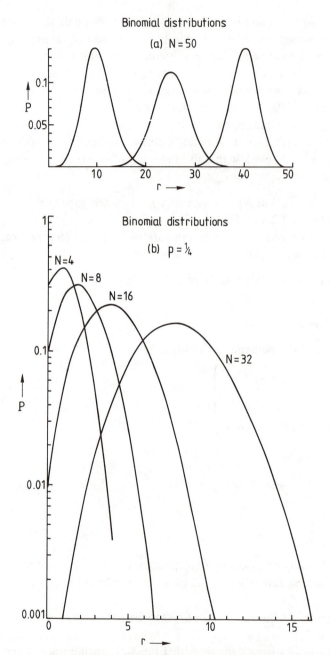

Fig. 3.1. Binomial distributions, characterised by the number of trials N and the probability of success p. The graphs show the probability P of obtaining r successes. (a) N is kept fixed at 50, and the three distributions correspond to $p = \frac{1}{5}, \frac{1}{2}$ and $\frac{4}{5}$ respectively. The distributions are of course defined only at integral r; the curves here and in (b) are drawn merely to link these points. The variance of a

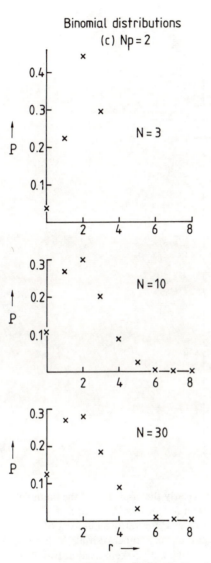

Binomial distributions
(c) Np = 2

binomial distribution is $Np(1-p)$ and hence the middle one is the widest (and thus has the lowest peak). The distributions for $p = \frac{1}{5}$ and $\frac{4}{5}$ are mirror images of each other about the line $r = 25$. (*b*) Here p is kept constant. The probability P is plotted on a logarithmic scale. As N becomes large, the distributions tend to the inverted parabola shape of a Gaussian distribution of mean Np and variance $Np(1-p)$. (*c*) Three binomial distributions in which N increases, but p correspondingly decreases such that Np is constant (and equals two). For large N, these tend to the Poisson distribution with mean and variance Np. For the $N = 30$ case, the binomial probabilities are barely distinguishable (on the scale shown) from the corresponding Poisson ones.

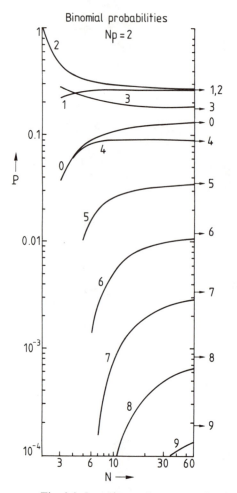

Fig. 3.2. In order to show more clearly the approach of the binomial distributions to that of the Poisson for the case Np fixed and N increasing, the binomial probabilities P for various numbers of successes are plotted as a function of the number of trials N; both scales are logarithmic. As in Fig. 3.1(c), Np is kept fixed at two. The values of r are shown by each curve, and the corresponding Poisson limits are denoted by arrows alongside the right-hand axis.

formula that applies to statistical errors in many situations involving the counting of independent events during a fixed interval.

As $\lambda \to \infty$, the Poisson distribution tends to a Gaussian one; for this approximation to be reasonably valid, it turns out that five or more is usually a good enough approximation to infinity for λ. This is very useful

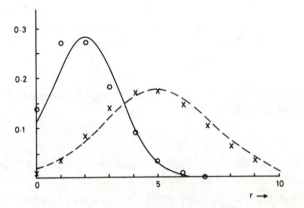

Fig. 3.3. Comparison of Poisson and Gaussian distributions when the mean is two (open circles for Poisson, solid curve for Gaussian) and five (crosses and dashed curve). In the latter case, agreement is quite good, and also the fractional area of the Gaussian with r negative is small. This is the basis of the rule of thumb that five events are sufficient to pretend that errors are Gaussian distributed.

since many statistical calculations are much simpler to perform if the errors are Gaussian distributed.

Comparisons are made between Poisson distributions with $\lambda = 2$ or 5 and the appropriate Gaussian distributions in Fig. 3.3. It is to be remembered that the Poisson distribution is defined only at non-negative integers while the Gaussian distribution is continuous and extends down to $-\infty$.

Examples of variables which are likely to be Poisson distributed are

Example (i)
The number of particles detected by a counter in a time t, in a situation where the particle flux ϕ and detector efficiency are independent of time, and where counter dead-time τ is such that $\phi\tau \ll 1$. (See Example (vii) below.)

Example (ii)
The number of interactions produced in a thin target when an intense pulse of N beam particles is incident on it.

Example (iii)
The number of entries in a given bin of a histogram when the data are accumulated over a fixed time interval.

It is important to remember that if the rate of the basic process changes (as a function of time or of position), then the observed distribution of events may not follow the Poisson distribution. Examples of this are

Example (iv)
The decay of a small amount of material over a period of time which is significant compared with the lifetime of the source.

Example (v)
The number of interactions produced by a beam consisting of a small number of particles, incident upon a thick target.

In these last two examples, the basic event rate decreases as a function of time or of position respectively. Furthermore, in both cases, there is a physical upper limit to the number of events of interest that can occur (the number of decays cannot exceed the initial number of nuclei in (iv); while in (v) the number of interactions is bounded by the number of beam tracks), while the Poisson distribution extends up to infinity.

Example (vi)
The number of people who die while operating computers each year is not Poisson distributed since although the probability of dying may remain constant, the number of people who operate computers increases from year to year.

Apart from a varying basic event rate, another feature which spoils the Poisson distribution is when the occurrence of individual events is not independent, as in the example below.

Example (vii)
A counter which is used to detect particles has a dead time of 1 μsec. This means that if a second particle passes within 1 μsec after one which has been recorded, the counter is incapable of recording the second particle. Then the detection of a particle is not independent of other particles which have passed through the counter. If the particle flux is low, this will not be serious, but no matter how high the flux is, this particular counter is incapable of recording more than 10^6 particles per second. Thus in high particle fluxes, the observed counting rate will not be Poisson distributed.

A relationship between Poisson and binominal distributions is demonstrated very nicely by considering an example of measuring a forward–backward asymmetry in an experiment which studies an angular distribution. We assume that there are B events observed in the backward and F in the forward hemisphere, making N events altogether. The probability P of observing these numbers is given by the product of the probability P_P obtaining N events altogether, and the probability P_B of observing F events in the forward hemisphere, given N events altogether; these two probabilities are obtained respectively from the Poisson distribution of mean v, and the binomial distribution when the probability of an individual event being in the forward hemisphere is f, i.e.

$$P = P_P P_B$$

$$= \left\{\frac{e^{-v} v^N}{N!}\right\} \times \left\{\frac{N!}{B!\,F!}\, f^F (1-f)^B\right\}. \tag{3.9}$$

After some trivial algebra, this reduces to

$$P = [e^{-vf} (vf)^F / F!] \times [e^{-v(1-f)} (v(1-f))^B / B!]. \tag{3.10}$$

Thus as well as regarding P as the product of a Poisson probability in N times a binomial probability in the number F conditional on the given N (eqn 3.9), we can instead think of P as the product of two independent Poisson distributions in the number of forward and the number of backward events separately (eqn 3.10).

We can repeat this procedure of dividing up the angular distribution as often as we want. Thus a histogram of the resulting angular distribution can be regarded as either (i) a Poisson distribution in the overall number of events N, and the corresponding (multinomial) distribution for obtaining $n_1, n_2, n_3, \ldots$ events in each bin, assuming the given total number of events N; or (ii) independent Poisson distributions in each bin of the histogram.

3.3 Gaussian distribution

Some simple properties of Gaussian distributions have already been discussed in Section 1.4.3 to which the reader is referred. For convenience, we here rewrite the Gaussian of mean μ and standard deviation σ as

$$P(x) = \frac{1}{\sqrt{(2\pi)}\,\sigma} \exp\{-(x-\mu)^2 / 2\sigma^2\}. \tag{3.11}$$

As previously stated, the Gaussian distribution is of wide applicability in the theory of errors. One very nice example of deriving such a distribution under very weak assumptions is provided by the inexperienced darts player. We can write the distribution of points where his darts hit the board either as a function of the horizontal and vertical co-ordinates

($f(x, y)$) or as a function of the polar co-ordinates with respect to the centre of the board ($g(r, \theta)$). We assume that f can be factorised into independent functions of x and y, and that g is independent of θ (i.e. we are neglecting the influence of gravity). Then

$$g(r) = f(x, y) = h(x) k(y). \tag{3.12}$$

Therefore

$$\frac{\partial g}{\partial \theta} = \frac{\partial f}{\partial x}\frac{\partial x}{\partial \theta} + \frac{\partial f}{\partial y}\frac{\partial y}{\partial \theta} = 0,$$

whence

$$\frac{h'(x)}{xh(x)} = \frac{k'(y)}{yk(y)} = A,$$

with solution

$$f(x, y) = C \exp(Ax^2) \exp(Ay^2) \tag{3.13}$$

or

$$g(r) = C \exp(Ar^2). \tag{3.13'}$$

Assuming that the darts player is sufficiently competent that more darts arrive near the origin than at an infinite distance from the board, then A must be chosen as negative, and the resulting distribution in x and y or in r is Gaussian.†

When using Gaussian distributions, we are often interested in integrals of the form

$$I_n = \int_{-\infty}^{+\infty} y^n\, e^{-y^2}\, dy, \tag{3.14}$$

and as a mathematical aside we give a mnemonic for deriving their values. For n even, we start with

$$I_0 = \int_{-\infty}^{+\infty} e^{-y^2}\, dy = \sqrt{\pi}. \tag{3.15}$$

Then

$$I_0(\alpha) = \int_{-\infty}^{+\infty} e^{-\alpha y^2}\, dy = \sqrt{(\pi/\alpha)}. \tag{3.15'}$$

---

† If we regard our friend as throwing darts not in ordinary space but in velocity space, and we allow the target to be 3-dimensional rather than the usual 2, the distribution of darts in velocity space gives us the well known Maxwell–Boltzmann factor $\exp(-V^2/V_0^2)$ of kinetic theory, where V_0^2 must be identified with $2kT/m$ to obtain the correct mean square velocity $\overline{V^2}$ of the molecules at temperature T (m is the mass of a molecule and k is Boltzmann's constant).

We then differentiate eqn (3.15′) as many times as is necessary with respect to α. For example,

$$\int_{-\infty}^{+\infty} (-y^2)\, e^{-\alpha y^2}\, dy = -I_2(\alpha) = -\frac{1}{2\alpha}\, \sqrt{(\pi/\alpha)} \tag{3.16}$$

from which I_2 is obtained by setting α equal to unity.

For odd n, I_n is zero. We thus define

$$J_n = \int_0^\infty y^n\, e^{-y^2}\, dy. \tag{3.14′}$$

Starting with

$$J_1 = \int_0^\infty y\, e^{-y^2}\, dy = \tfrac{1}{2}, \tag{3.17}$$

we obtain

$$J_1(\alpha) = \int_0^\infty y\, e^{-\alpha y^2}\, dy = \frac{1}{2\alpha}, \tag{3.17′}$$

from which all the odd J_n can be obtained by differentiation.

The properties of Gaussian distributions are commonly used in interpreting the significance of experimental results. Their use in this context relies on the following assumptions:

(i) The value of the quantity of interest has been correctly calculated (e.g. there are no important systematic biasses)

(ii) The magnitude of the error has been correctly calculated. This is particularly important, in that an incorrect estimate of the accuracy of the experiment could have a very large effect on the calculated significance of our result and hence on our conclusions. (See also the remarks about *estimates* of σ in the paragraph below eqn (3.19)). Thus an underestimate of our experimental errors by a factor of two could change a two standard deviation effect (which occurs at the 5% level) into a four standard deviation one (whose probability is only 6.10^{-5} i.e. such an effect 'cannot' happen if the theory is correct).

(iii) The form of the experimental resolution is such that the Gaussian approximation is reasonable. This is almost always untrue, in that the probability of obtaining large deviations ($\gtrsim 3\sigma$) from the correct value is often larger than given by the Gaussian distribution. This effect, which has already been discussed at the end of Section 1.4.3, is also likely to result in an artificial

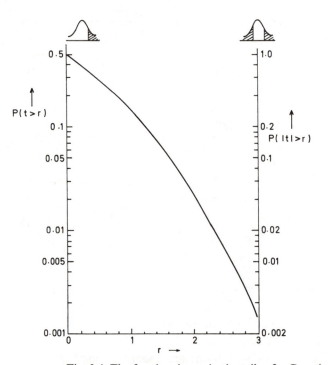

Fig. 3.4. The fractional area in the tails of a Gaussian distribution, i.e. the area with t greater than some specified value r, where t is the distance from the mean, measured in units of the standard deviation. The scale on the left-hand vertical axis refers to the one-sided tail, while the right-hand one is for $|t|$ larger than r. Thus for $r = 0$, the probabilities are $\frac{1}{2}$ and 1 respectively.

enhancement of our estimation of the significance of observed deviations.

Let us assume that the above complications are absent, and consider a simple example of this type of problem. We measure the lifetime of the neutron in our experiment as 950 ± 20 secs. A certain theory predicts that the lifetime is 910 secs. To what extent are these numbers in agreement?†

We consult Fig. 3.4, which is a graph showing the fractional area under the Gaussian curve with

$$|t| > r, \tag{3.18}$$

where

$$t = \frac{x - \mu}{\sigma}, \tag{3.19}$$

† This is in fact a simple example of 'hypothesis testing' to which we return in Section 4.6.

i.e. it gives (on the right-hand vertical scale) the area in the tails of the Gaussian beyond any value r of the parameter t, which is plotted on the horizontal axis. In our example of the neutron lifetime, $t = 2$ and the corresponding probability is 4.6%. Thus if 1000 experiments of the same precision as ours were performed to measure the neutron lifetime, and if our theory is correct, *and* if the experiments are bias-free, then we expect about 46 of them to differ from the predicted value by at least as much as ours. Of course, we still have to make up our mind whether we regard the theory (and the experiment) as satisfactory or not, but at least we have a number on which to base our judgement.

It is important to realise that σ in eqn (3.19) is supposed to be the true value of the experimental resolution. In some cases, we will simply estimate this from the observed spread of a repeated set of measurements (see Section 1.4.2). We then expect fluctuations in the denominator of t to widen its distribution as compared with a Gaussian; the expected distribution is Student's t, and depends on the number of observations N used to estimate σ. Since such information about other people's experiments is often unobtainable, most physicists would use a Gaussian distribution to estimate the significance of t. The difference between these distributions can be important for small N. Thus the probability of obtaining $|t| > 3$ is 0.3% for a Gaussian distribution, but is 3% for Student's t with $N = 6$, and 20% for $N = 2$.

In most cases, the theoretical estimate y_{th} will have an uncertainty σ' associated with it; theory, after all, is based on experiment, and hence predictions in general are calculated from other experimental numbers which of course have their own experimental errors. In that case, we repeat the above procedure of consulting Fig. 3.4, but we redefine t for this case as

$$t = \frac{y_{obs} - y_{th}}{\sqrt{(\sigma^2 + \sigma'^2)}}, \tag{3.19'}$$

where our measured value is $y_{obs} \pm \sigma$. The denominator of (3.19') arises because it is the error on the numerator, assuming that the errors on y_{obs} and y_{th} are uncorrelated (see Section 1.5.1).

Sometimes we are interested in the sign of possible deviations from predicted values.

Example (i)

Is there any evidence for other processes, which we have not included in our theory, contributing incoherently to the neutron decay? In other words, is the observed lifetime for neutron decay *smaller* than the predicted value?

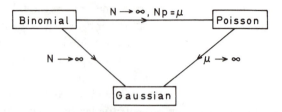

Fig. 3.5. The relationships among the Binomial, Poisson and Gaussian distributions. *N* and *p* are respectively the number of trials and the probablity of success in each trial in the Binomial case. The mean of the Poisson distribution is denoted by μ. In the particular case of the Gaussian derived as the limit of the Poisson distribution for large μ, the mean and variance of the distribution are equal.

Example (ii)

A nuclear reactor will explode if the neutron production rate is greater than a certain value λ_c. We measure the rate as $\lambda \pm \sigma$ where λ is a bit smaller than λ_c. Since we are crucially interested in the danger of explosion, we want to know what is the probability that the true rate is *greater* than λ_c.

In cases where the sign of the possible deviation is of significance, we use the left-hand scale of Fig. 3.4, which gives the area in the Gaussian with $t > r$ (or, by symmetry, with $t < -r$), i.e. it is the area in one of the tails.

Fig. 3.4 shows that measured values too far from the predicted value are unlikely, but so are those that are too close. For example, if we measure the neutron lifetime as 909 ± 200 secs, we are suspiciously close to the prediction of 910 seconds. From Fig. 3.4 it can be seen (in principle rather than in practice since the scales are not optimal for very small values of t) that the probability of being within $\frac{1}{200}$ of a standard deviation from a specific value is only 0.4%. Thus

  (i) we are unusually lucky on this occasion; or
  (ii) our error is over-estimated; or
  (iii) we in fact knew the predicted value before we started the experiment, and (perhaps unconsciously) cheated to get close to the 'right' answer.

The relations among the Binomial, Poisson and Gaussian distributions are demonstrated in Fig. 3.5.

The Gaussian distribution has already been compared with the corresponding Poisson distribution in Fig. 3.3. Fig. 3.6 shows a Gaussian and a simple Breit–Wigner distribution drawn on the same axes. Narrow nuclear or particle states which are broadened significantly by experimental

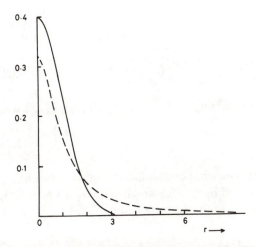

Fig. 3.6. Comparison of a Gaussian distribution of mean zero and unit variance (solid curve) with a Breit–Wigner distribution (dashed curve) defined as $\dfrac{1}{\pi}\dfrac{1}{r^2+1}$. Both curves are symmetric about $r = 0$, but the Breit–Wigner has a very long tail.

resolution will tend to have a Gaussian shape, while wide states for which the experimental resolution is less important may be more nearly Breit–Wigner shaped.

3.4 Gaussian distribution in two variables

(a) Uncorrelated variables

In this and the following section, we transform our variables, so that the distribution is a maximum at the origin; this simplifies the appearance of the algebra. The formulation below applies to any Gaussian distribution. We concentrate, however, on that aspect where we have a pair of measurements x and y of two different experimental quantities, each of which has an uncertainty associated with it arising from the finite experimental resolution. The aim of this section is to familiarise the reader with the error matrix without introducing any essentially new concepts.

For our two variables, we have probability distributions given by

$$P(x) = \frac{1}{\sqrt{(2\pi)}\,\sigma_x} \exp\left(-\frac{1}{2}\frac{x^2}{\sigma_x^{\;2}}\right)$$

and

$$P(y) = \frac{1}{\sqrt{(2\pi)}\,\sigma_y} \exp\left(-\frac{1}{2}\frac{y^2}{\sigma_y^{\;2}}\right).$$

(3.20)

Now if x and y are independent,† from formula (2.6) we obtain

$$P(x, y) = P(x) P(y)$$

$$= \frac{1}{2\pi \sigma_x \sigma_y} \exp\left(-\frac{1}{2}\left(\frac{x^2}{\sigma_x^2} + \frac{y^2}{\sigma_y^2}\right)\right). \tag{3.21}$$

This will be down on the probability at the origin by a factor of $\sqrt{e}$ when

$$\frac{x^2}{\sigma_x^2} + \frac{y^2}{\sigma_y^2} = 1. \tag{3.21'}$$

(This is to be compared with the corresponding fact that in the 1-dimensional case, the probability is reduced by this factor when $x = \pm\sigma$.) Now, if we really want to, we can rewrite (3.21') in matrix notation as

$$(x \quad y)\begin{pmatrix} \frac{1}{\sigma_x^2} & 0 \\ 0 & \frac{1}{\sigma_y^2} \end{pmatrix}\begin{pmatrix} x \\ y \end{pmatrix} = 1. \tag{3.21''}$$

Finally we can invert the 2×2 matrix in the above equation to obtain the matrix

$$\begin{pmatrix} \sigma_x^2 & 0 \\ 0 & \sigma_y^2 \end{pmatrix} \tag{3.22}$$

which is known as the error matrix for x and y. The diagonal terms σ_x^2 and σ_y^2 are respectively the variances of x and of y, while the off-diagonal zeroes indicate that the errors of x and y are uncorrelated. The 2×2 matrix in (3.21'') is known as the inverse error matrix. In such a simple example, nothing is gained by using matrix techniques; their main use is for dealing with correlations.

In general the element M_{ij} of an error matrix for a set of variables $x_1 x_2, ..., x_n$ is defined as the expectation value

$$\langle (x_i - \bar{x}_i)(x_j - \bar{x}_j) \rangle. \tag{3.23}$$

Hence we see that the error matrix M is symmetric and the off-diagonal terms are $\mathrm{cov}(x_i, x_j)$. Also we can use (3.23) to evaluate the error matrix in cases where the distributions (3.20) or (3.21) are not Gaussian. (This corresponds to the fact that the variance σ^2 can be defined for (most) 1-dimensional distributions, but has many especially simple properties for the Gaussian distribution. For the 2-dimensional case, we find it

† An extreme counter-example is if $y = x$, when $P(x, y)$ is zero unless $x = y$.

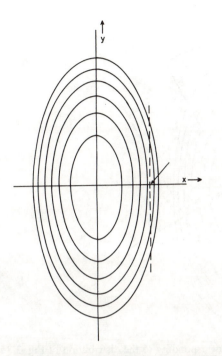

Fig. 3.7. Contour lines for a Gaussian distribution in two dimensions, the variables x and y being uncorrelated. The variables have been transformed by a simple translation so that the distribution peaks at the origin. The probability distribution is given by formula (3.21); the contours of constant probability are ellipses. The dashed line corresponds to a choice of a non-zero value for x; inspection of the contour lines (or of formula (3.21)) shows that the maximum value of the distribution on this line occurs at the point denoted by the arrow, i.e. the maximum of the distribution for any value of x occurs at a value of y (in fact, zero) which is independent of the specific choice of the x-value.

advantageous to use the Gaussian distribution to introduce the idea of the error matrix, which is, however, of much wider applicability.)

Some of the contours of constant $P(x, y)$ of formula (3.21) are shown in Fig. 3.7. We see that if we select any non-zero value of x, then the value of y which maximises $P(x, y)$ is still zero. This is consistent with the fact that x and y are uncorrelated. (It is not a proof, since we have not demonstrated that the shape of the whole distribution is unaffected by our choice of x.)

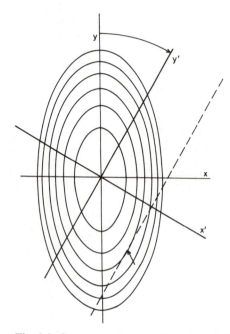

Fig. 3.8. Contours corresponding to the distribution of Fig. 3.7 in the
xy system, where the distribution is uncorrelated. The maximum of
the distribution is still at the origin. We then rotate the axes clockwise
by an angle θ, to the new $x'y'$ axes. As viewed in the new axes (i.e.
rotate the book anticlockwise by 30°), the distribution is seen to be
correlated in the $x'y'$ variables. Thus if we choose a non-zero value of
x' (the dashed line), the maximum of the distribution on this line is at
the point denoted by the arrow. That is, the best choice of y' is
dependent on the particular value of x'; this is the correlation.

As a specific example of the error matrix (3.22) we could have

$$\left.\begin{aligned} \sigma_x &= \frac{\sqrt{2}}{4} = 0.354, \\[2mm] \sigma_y &= \frac{\sqrt{2}}{2} = 0.707. \end{aligned}\right\} \tag{3.24}$$

Then

$$P(x, y) = P(0, 0)/\sqrt{e}$$

when

$$8x^2 + 2y^2 = 1. \tag{3.25}$$

We use this example below.

(b) *Correlated variables*

The simplest way to introduce correlations is to start with the uncorrelated case which we now understand, and then to rotate the axes clockwise by an angle θ (see Fig. 3.8). Then the variables x' and y' with respect to the new axes are related to the old ones by

$$\left.\begin{array}{l} x' = x \cos\theta - y \sin\theta, \\ y' = x \sin\theta + y \cos\theta. \end{array}\right\} \tag{3.26}$$

If we choose, say, θ as 30°, then our ellipse (3.25) becomes

$$\tfrac{1}{2}[13x'^2 + 6\sqrt{3}\,x'y' + 7y'^2] = 1. \tag{3.27}$$

We stress that the probability distribution is the same as in the example above and hence (3.27) still is the contour corresponding to a reduction in $P(x', y')$ by $\sqrt{e}$. But because we have rotated our axes we have produced the cross-term $6\sqrt{3}\,x'y'$ in (3.27) and it is this which is indicative of the correlation between the x'- and y'-errors.

As in the uncorrelated case, we can rewrite (3.27) in matrix form:

$$(x'\ y')\begin{pmatrix} \dfrac{13}{2} & \dfrac{3\sqrt{3}}{2} \\[2mm] \dfrac{3\sqrt{3}}{2} & \dfrac{7}{2} \end{pmatrix}\begin{pmatrix} x' \\ y' \end{pmatrix} = 1. \tag{3.27'}$$

If we invert the central matrix, we obtain the error matrix as

$$\frac{2}{64}\begin{pmatrix} 7 & -3\sqrt{3} \\ -3\sqrt{3} & 13 \end{pmatrix}. \tag{3.28}$$

Then

$$\left.\begin{array}{l} \sigma_{x'}^2 = \dfrac{14}{64} = (0.468)^2, \\[3mm] \sigma_{y'}^2 = \dfrac{26}{64} = (0.637)^2, \\[3mm] \mathrm{cov}\,(x', y') = -\dfrac{6\sqrt{3}}{64} = -(0.403)^2. \end{array}\right\} \tag{3.29}$$

The negative sign of $\mathrm{cov}\,(x', y')$ means that as x' *increases* from zero, the value of y' that maximises $P(x', y')$ (for that particular value of x') *decreases*. These values (3.29) are to be compared with the previous ones

$$\left.\begin{array}{l} \sigma_x^2 = (0.354)^2, \\[2mm] \sigma_y^2 = (0.707)^2, \\[2mm] \mathrm{cov}\,(x, y) = 0. \end{array}\right\} \tag{3.30}$$

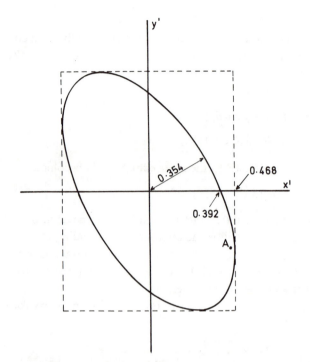

Fig. 3.9. Diagram of correlated Gaussian variables, demonstrating the features of the ellipse (3.27), which is the contour on which $P(x', y')$ is smaller than $P(0, 0)$ by a factor of $e^{\frac{1}{2}}$ (i.e. points within the ellipse correspond to x', y' combinations which are 'quite likely', while those outside are less so). The semi-axes of the ellipse (one of which is of length 0.354) are defined in terms of the error matrix eigenvalues, the dimensions of the rectangle enclosing the ellipse are related to the diagonal elements of the error matrix, while the intersections of the ellipse with the axes (one of which is 0.392) are calculated from the diagonal elements of the inverse error matrix. For a point such as A within the ellipse, the value of x' is greater than 0.392; hence the intersections of the ellipse with the axes give an underestimate of the range of likely values of that particular parameter. It is the diagonal elements of the error matrix (rather than those of the inverse error matrix) which thus describe the likely range of values of a parameter. The complete area within the rectangle, however, is an overestimate of likely (x', y') combinations.

The covariance term of the error matrix is negative since the major axis of the ellipse has a negative gradient; for x' constant and *negative*, the maximum of $P(x', y')$ occurs at *positive* y'.

The ellipse (3.27) is redrawn in Fig. 3.9 with respect to x'- and y'-axes. The significant parameters are given in terms of the error matrix as follows.

(i) The *semi-axes* of the ellipse are given by the square roots of the *eigenvalues of the error matrix.*

(ii) The *maximum* value of x' on the ellipse is given by the square root of the first *diagonal element of the error matrix.*

(iii) The ellipse *intersects* the x'-axis at a value given by the square root of the reciprocal of the first *diagonal element of the inverse error matrix.* (Note that 0.392 is equal to $\sqrt{\frac{2}{13}}$.)

Thus if we use the interior of the contour (3.27) to indicate the likely values of the parameters x' and y', then it is the number as deduced in (ii) above (i.e. that from the diagonal elements of the error matrix) which gives the likely range of the x'-values. We can similarly deduce the likely y'-range from the other diagonal element of the error matrix. The likely pairs of (x', y')-values are confined to a region smaller than the rectangle defined by the maximum x'- and y'-values separately, and are bounded by the ellipse which, because of the correlations, is tilted with respect to the axes.

In some methods involving more than one variable, it is easier to estimate the diagonal elements of the inverse error matrix than to obtain the diagonal elements of the error matrix itself. The parameters (iii) deduced from the inverse error matrix underestimate the range of likely values of each parameter separately; as can immediately be seen from Fig. 3.9 there are points within the ellipse with have x'-values larger than that of the intersection of the ellipse with the x'-axis. The effect becomes most serious when the correlations are strongest.

The general form of a Gaussian distribution in two variables (including correlations) can be written as

$$P(x, y) = \frac{1}{2\pi \sigma_x \sigma_y} \frac{1}{\sqrt{(1-\rho^2)}} \exp \left\{ -\frac{1}{2} \frac{1}{1-\rho^2} \left(\frac{x^2}{\sigma_x^2} + \frac{y^2}{\sigma_y^2} - \frac{2\rho xy}{\sigma_x \sigma_y} \right) \right\},$$
(3.31)

where

$$\rho = \frac{\text{cov}(x, y)}{\sigma_x \sigma_y}.$$
(3.32)

The parameter ρ is known as the correlation coefficient for x and y and is such that

$$|\rho| \leqslant 1.$$
(3.33)

Not surprisingly, the correlation coefficient is zero when the variables are uncorrelated, in which case (3.31) reduces to (3.21).

The probability distribution of y for a specific value X of x (i.e. the conditional probability $P(y/X)$) is also Gaussian distributed. Its mean is

$$\bar{y} + \rho \frac{\sigma_y}{\sigma_x} (X - \bar{x}) \tag{3.34}$$

and its variance is

$$\sigma_y^2 (1 - \rho^2), \tag{3.35}$$

where we have deliberately reintroduced the values of the means $\bar{x}$ and $\bar{y}$ of the overall Gaussian distributions which earlier we had set to zero. Formula (3.34) explicitly demonstrates the correlation; the mean of the y distribution differs from $\bar{y}$ by an amount which is proportional to the correlation coefficient and to the offset of X from the mean value $\bar{x}$. We also see that the variance of the conditional probability distribution is smaller than that of y for the overall distribution (σ_y^2). This corresponds to the general principle that extra information (the particular value X for the variable x) often increases the precision of the answer.

If we want to extend (3.31) to K variables, it becomes

$$P = \frac{1}{(2\pi)^{K/2}} \frac{1}{|M|^{\frac{1}{2}}} \exp\left(-\tfrac{1}{2}\tilde{\mathbf{x}} \mathbf{M}^{-1} \mathbf{x}\right), \tag{3.36}$$

where $\mathbf{M}$ is the error matrix, $\mathbf{x}$ is the vector of variables (and $\tilde{\mathbf{x}}$ its transpose) and $|M|$ is the determinant of the matrix $\mathbf{M}$. For the two variable case

$$\mathbf{M} = \begin{pmatrix} \sigma_x^2 & \mathrm{cov}\,(x, y) \\ \mathrm{cov}\,(x, y) & \sigma_y^2 \end{pmatrix} \tag{3.37}$$

and

$$\mathbf{M}^{-1} = \frac{1}{1 - \rho^2} \begin{pmatrix} \dfrac{1}{\sigma_x^2} & -\dfrac{\rho}{\sigma_x \sigma_y} \\ -\dfrac{\rho}{\sigma_x \sigma_y} & \dfrac{1}{\sigma_y^2} \end{pmatrix}. \tag{3.38}$$

In order to appreciate the significance of the off-diagonal terms, it is useful to consider the situation where σ_x^2 and σ_y^2 are kept fixed but $\mathrm{cov}\,(x, y)$ (or ρ – see eqn (3.32)) is varied. Then all the ellipses are enclosed by the same rectangle, but the shape of the ellipse is determined by ρ. (See Fig. 3.10.)

We have constantly emphasised how important it is to present error

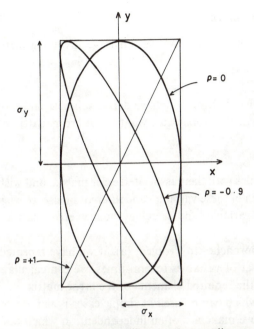

Fig. 3.10. Some ellipses, corresponding to formula (3.31) with $P(x, y)$ set equal to $P(0, 0)/\sqrt{e}$, with σ_x and σ_y constant, but for various values of ρ. When $\rho = 0$, the axes of the ellipse are parallel to the x- and y-axes, and the variables are uncorrelated. For $\rho = +1$, the ellipse degenerates to a straight line of gradient σ_y/σ_x. The ellipse for $\rho = -0.9$ is thin in shape, and the major axis has a negative gradient. All the ellipses touch the same rectangle, which has sides of length $2\sigma_x$ and $2\sigma_y$.

estimates on all measured quantities. When several variables are measured simultaneously, it is the error matrix which provides the necessary information on the magnitudes of the errors and on their correlations (if any). But correlations, especially in many variables, can become a very messy subject – it is tiresome to estimate all the correlations, and it is difficult to assess visually or mentally the significance of a whole error matrix. Thus it is highly desirable where possible to choose the variables in a multi-dimensional problem intelligently, i.e. to find a set of variables whose errors are Gaussian and uncorrelated. But this is often easier said than done.

3.5 Using the error matrix

There are two sorts of problems that we can envisage involving error matrices.

Problem (a)
We make a set of measurements from which we derive some variables of interest to us. For our particular apparatus, what is the error matrix of our variables?

Problem (b)
We are given a set of variables with their associated error matrix, and wish to calculate either a function of these variables and its error or else we wish to transform to some new variables and calculate the new error matrix.

There are basically two approaches to Problem (a). If we have repeated the measurements of our set of variables several times, we can calculate from the spread of values the required elements of the error matrix.

An alternative method which can be used in simple cases is as follows. The actual measurements we make are often independent, in which case the off-diagonal elements of their error matrix are all zero. By simple considerations concerning each of the measurements separately, we may then deduce the diagonal elements of their error matrix, and hence we now have the whole error matrix. The variables of interest are calculated from the measurements in terms of some relatively simple transformation. Then we can use the method described in (ii) below to transform from the original diagonal error matrix of the measurements to the error matrix of the variables. Such procedures are carried out in Examples (iii) and (iv) of Section 3.6.

The above methods correspond exactly to determining the variance on a single measurement either by observing the spread on a series of measurements, or by considering the intrinsic accuracy of the apparatus and deducing how well our variable can be determined (see Section 1.3).

We thus pass on to Problem (b) of manipulations with the error matrix.

Situation (i). A function of the variables
We are given a pair of variables x_a and x_b and their error matrix. We are required to calculate a specific function

$$y = y(x_a, x_b) \tag{3.39}$$

and we want to know what is the error on y.

As in the corresponding simple problems of Section 1.5, we first differentiate, then square the resulting expression and finally take the expectation value, since we want to obtain the value of $\langle (y-\bar{y})^2 \rangle = \overline{\delta y^2}$.

$$\overline{\delta y^2} = \left(\frac{\partial y}{\partial x_a}\right)^2 \overline{\delta x_a{}^2} + \left(\frac{\partial y}{\partial x_b}\right)^2 \overline{\delta x_b{}^2} + 2 \frac{\partial y}{\partial x_a} \frac{\partial y}{\partial x_b} \overline{\delta x_a \delta x_b}. \tag{3.40}$$

This can be written in matrix form as

$$\overline{\delta y^2} = \left(\frac{\partial y}{\partial x_a} \frac{\partial y}{\partial x_b}\right) \begin{pmatrix} \overline{\delta x_a{}^2} & \overline{\delta x_a \delta x_b} \\ \overline{\delta x_a \delta x_b} & \overline{\delta x_b{}^2} \end{pmatrix} \begin{pmatrix} \dfrac{\partial y}{\partial x_a} \\ \dfrac{\partial y}{\partial x_b} \end{pmatrix} \tag{3.40'}$$

The 2×2 matrix in (3.40') is simply the error matrix M of x_a and x_b, and the vector is the derivative vector D for the transformation (3.39). Thus we finally write the result for the error on a function of some variables as

$$\sigma_y{}^2 = \tilde{D} M D. \tag{3.41}$$

We must of course remember that in obtaining the above result, we had to differentiate. Hence if the function (3.39) is not linear, the errors had better be small or formula (3.41) will not be accurate.

Situation (ii). Change of variables

We again are provided with x_a and x_b and their corresponding error matrix, but this time we wish to change to new variables p_i and p_j, and to calculate the new error matrix. The transformation is given by

$$\left.\begin{aligned} p_i &= p_i(x_a, x_b), \\ p_j &= p_j(x_a, x_b). \end{aligned}\right\} \tag{3.42}$$

As usual we start by differentiating, and obtain

$$\delta p_i = \frac{\partial p_i}{\partial x_a} \delta x_a + \frac{\partial p_i}{\partial x_b} \delta x_b, \tag{3.43}$$

and similarly for p_j. Then we square this expression and take its expectation value, to obtain

$$\overline{\delta p_i{}^2} = \left(\frac{\partial p_i}{\partial x_a}\right)^2 \overline{\delta x_a{}^2} + \left(\frac{\partial p_i}{\partial x_b}\right)^2 \overline{\delta x_b{}^2} + 2 \left(\frac{\partial p_i}{\partial x_a}\right)\left(\frac{\partial p_i}{\partial x_b}\right) \overline{\delta x_a \delta x_b}. \tag{3.44}$$

We can also obtain an analogous expression for $\overline{\delta p_j{}^2}$. Finally by multiplying (3.43) by the corresponding expression for δp_j, we deduce

$$\overline{\delta p_i \delta p_j} = \left(\frac{\partial p_i}{\partial x_a} \frac{\partial p_j}{\partial x_a}\right) \overline{\delta x_a{}^2} + \left(\frac{\partial p_i}{\partial x_b} \frac{\partial p_j}{\partial x_b}\right) \overline{\delta x_b{}^2}$$

$$+ \left(\frac{\partial p_i}{\partial x_a} \frac{\partial p_j}{\partial x_b} + \frac{\partial p_i}{\partial x_b} \frac{\partial p_j}{\partial x_a}\right) \overline{\delta x_a \delta x_b}. \tag{3.45}$$

The expressions for $\overline{\delta p_i{}^2}$, $\overline{\delta p_j{}^2}$ and $\overline{\delta p_i \delta p_j}$ can be compressed into the single matrix equation

$$
\begin{pmatrix} \overline{\delta p_i{}^2} & \overline{\delta p_i \delta p_j} \\ \\ \overline{\delta p_i \delta p_j} & \overline{\delta p_j{}^2} \end{pmatrix} = \begin{pmatrix} \dfrac{\partial p_i}{\partial x_a} & \dfrac{\partial p_i}{\partial x_b} \\ \\ \dfrac{\partial p_j}{\partial x_a} & \dfrac{\partial p_j}{\partial x_b} \end{pmatrix} \begin{pmatrix} \overline{\delta x_a{}^2} & \overline{\delta x_a \delta x_b} \\ \\ \overline{\delta x_a \delta x_b} & \overline{\delta x_b{}^2} \end{pmatrix} \begin{pmatrix} \dfrac{\partial p_i}{\partial x_a} & \dfrac{\partial p_j}{\partial x_a} \\ \\ \dfrac{\partial p_i}{\partial x_b} & \dfrac{\partial p_j}{\partial x_b} \end{pmatrix} \cdot \quad (3.46)
$$

↑	↑	↑	↑
New error matrix	Transpose of T	Old error matrix	Transformation matrix T

N.B. It is essential to have the non-symmetric matrix T the right way round. If T and its transpose are interchanged, the wrong answer is obtained in all but the simplest of cases.

Situation (iii). Function of changed variables

We wish to calculate the error on a function

$$y = y(p_i, p_j), \qquad (3.47)$$

where the variables p_i and p_j are defined in terms of other variables x_a and x_b (whose error matrix M we know) by

$$\left. \begin{aligned} p_i &= p_i(x_a, x_b), \\ p_j &= p_j(x_a, x_b). \end{aligned} \right\} \qquad (3.42)$$

The result is simply obtained by combining our two previous situations, to give

$$\sigma_y{}^2 = \tilde{D}\tilde{T}MTD, \qquad (3.48)$$

where T is the transformation matrix appearing in eqn (3.46) and D is the derivative vector

$$\begin{pmatrix} \dfrac{\partial y}{\partial p_i} \\ \\ \dfrac{\partial y}{\partial p_j} \end{pmatrix}.$$

The big advantage of using matrices in such problems is that we are relieved of the obligation to think about how to solve any specific problem; and computers will merrily multiply matrices for almost as long as we want, without their becoming bored. (To do more than a couple of error matrix manipulations without the aid of a computer is a daunting prospect for even the most enthusiastic of data analysts.)

An example of the use of this type of matrix product is provided by a situation in which two charged particles are produced in an interaction, and then pass through a magnetic field. The co-ordinates of each track are recorded at several positions; the error matrix M_1 for these co-ordinate measurements is known (and may well be diagonal). We are interested in calculating some kinematic quantity associated with these tracks; it could, for example, be the mass of a resonant state which decays into our observed charged tracks. We first transform the co-ordinate measurements to momentum vectors at the interaction point. The new error matrix M_2 is obtained by using the transformation matrix, and is likely to contain off-diagonal terms corresponding to correlations between each particle's production angle in the bending plane and the magnitude of its momentum. Finally we calculate the derivative matrix (from the known functional dependence of the mass in terms of the momentum vectors) and use it with the error matrix M_2 to obtain the error on the mass. In a typical experiment we will observe many interactions of the above type, and it may well be necessary to calculate the error on the mass separately for each interaction. This is thus clearly a situation where the use of a computer is essential. (But in this experiment we would already be using a computer anyway in order to calculate the observed mass in each interaction.)

3.6 Specific examples of error matrix manipulations

Example (i). Very simple function
As a somewhat artificial but nevertheless instructive example, we consider the function

$$y = x + 2x \tag{3.49}$$

where the statistical error on the measurement of x is ε_x. Then the error matrix of the variables x and $2x$ is

$$M = \begin{pmatrix} \varepsilon_x^2 & 2\varepsilon_x^2 \\ 2\varepsilon_x^2 & 4\varepsilon_x^2 \end{pmatrix}, \tag{3.50}$$

where the derivation of the diagonal terms is completely trivial, and for the off-diagonal term is hopefully obvious.

The derivative matrix

$$D = \begin{pmatrix} \dfrac{\partial y}{\partial x} \\ \dfrac{\partial y}{\partial (2x)} \end{pmatrix} = \begin{pmatrix} 1 \\ 1 \end{pmatrix}, \tag{3.51}$$

and hence from (3.41)

$$\sigma_y{}^2 = \varepsilon_x{}^2 (1 \quad 1) \begin{pmatrix} 1 & 2 \\ 2 & 4 \end{pmatrix} \begin{pmatrix} 1 \\ 1 \end{pmatrix}$$

$$= 9\varepsilon_x{}^2. \tag{3.52}$$

Thus
$$\sigma_y = 3\varepsilon_x,$$

as expected. (Had we omitted the off-diagonal correlation terms, then we would have obtained the incorrect answer of $\sigma_y = \sqrt{5}\,\varepsilon_x$.)

An alternative derivation of the answer is obtained by regarding y as a function of two variables x and x rather than x and $2x$ as above. Then

$$M = \begin{pmatrix} \varepsilon_x{}^2 & \varepsilon_x{}^2 \\ \varepsilon_x{}^2 & \varepsilon_x{}^2 \end{pmatrix} \tag{3.50'}$$

and
$$D = \begin{pmatrix} 1 \\ 2 \end{pmatrix}, \tag{3.51'}$$

giving
$$\sigma_y{}^2 = \varepsilon_x{}^2 (1 \quad 2) \begin{pmatrix} 1 & 1 \\ 1 & 1 \end{pmatrix} \begin{pmatrix} 1 \\ 2 \end{pmatrix}$$

$$= 9\varepsilon_x{}^2, \tag{3.52'}$$

as before.

Finally we can rewrite (3.49) as

$$y = 3x \tag{3.49''}$$

and obtain $\sigma_y{}^2$ by the 'matrix' product (3.41) (where now all the 'matrices' are of size 1×1)

$$\sigma_y{}^2 = (3)(\varepsilon_x{}^2)(3)$$

$$= 9\varepsilon_x{}^2, \tag{3.52''}$$

yet again.

Example (ii). More realistic function

As a more realistic example, we calculate the error on the forward–backward asymmetry of an angular distribution

$$A = \frac{F-B}{F+B} = \frac{F-B}{N}, \tag{3.53}$$

where F and B are respectively the observed numbers of forward and backward produced particles in these interactions and N is their sum. If the errors σ_F and σ_B on F and B are uncorrelated,† then

$$\overline{\delta A^2} = \begin{pmatrix} \dfrac{\partial A}{\partial F} & \dfrac{\partial A}{\partial B} \end{pmatrix} \begin{pmatrix} \sigma_F{}^2 & 0 \\ 0 & \sigma_B{}^2 \end{pmatrix} \begin{pmatrix} \dfrac{\partial A}{\partial F} \\ \dfrac{\partial A}{\partial B} \end{pmatrix},$$

whence
$$\sigma_A = \frac{2FB}{N^2} \sqrt{\left[\left(\frac{\sigma_F}{F} \right)^2 + \left(\frac{\sigma_B}{B} \right)^2 \right]}. \tag{3.54}$$

If the numbers of observed interactions are Poisson distributed (i.e. $\sigma_F{}^2$ is equal to F, and similarly for B), then (3.54) reduces to

$$\sigma_A = \frac{1 - A^2}{2} \sqrt{\left(\frac{1}{F} + \frac{1}{B} \right)}. \tag{3.55}$$

Two extreme cases are:

(a) $F \sim B \sim \dfrac{N}{2}$ and the asymmetry $A \sim 0$.

Then $\quad \sigma_A \sim \dfrac{\delta N}{N}.$ \hfill (3.56)

For example, if 100 ± 10 events are observed in each of the two hemispheres, then

$$A = 0.00 \pm 0.07. \tag{3.57}$$

(b) $F \gg B$ and hence $A \sim +1$.

Then $\quad \sigma_A \sim \dfrac{2\delta B}{N}.$ \hfill (3.58)

Thus the error on A is dominated by the uncertainty on the smaller number of events. As an example, with $F = 991 \pm 32$ and $B = 9 \pm 3$

$$A = 0.982 \pm 0.006. \tag{3.59}$$

We assumed above that the numbers of events in each hemisphere are given by independent Poisson distributions, with total number N being given by $F + B$ and being variable. Alternatively, we can regard N as being fixed and F and B being given by the relevant binomial distribution. If p

---

† This could well be the case provided that N is not fixed at some predetermined value and also that measurement uncertainties are not so large that some particles cannot uniquely be defined as being in a specific hemisphere.

is the probability of a particle being produced in the forward hemisphere, then

$$\sigma_F{}^2 = \sigma_B{}^2 = Np(1-p)$$

$$\sim FB/N,$$

and since F and B are completely anti-correlated,

Since $$\mathrm{cov}\,(F, B) = -\sigma_F{}^2.$$

$$A = (F - B)/N,$$

$$\frac{\partial A}{\partial F} = -\frac{\partial A}{\partial B} = \frac{1}{N},$$

and

$$\sigma_A{}^2 = \left(\frac{1}{N} \quad -\frac{1}{N}\right)\left(\begin{matrix} \sigma_F{}^2 & -\sigma_F{}^2 \\ -\sigma_F{}^2 & \sigma_F{}^2 \end{matrix}\right)\left(\begin{matrix} \dfrac{1}{N} \\ -\dfrac{1}{N} \end{matrix}\right)$$

$$= \frac{4}{N^2}\frac{FB}{N},$$

which is identical to (3.55). This is not too surprising in view of the already noted equivalence of the two ways of looking at a distribution of forward and backward events (see Section 3.2).

An even simpler method for determining the error on the asymmetry for the fixed N case is to write

$$A = 1 - \frac{2B}{N}.$$

Since

$$\sigma_B = \sqrt{(FB/N)}$$

we immediately obtain

$$\sigma_A = \frac{2}{N}\sqrt{\left(\frac{FB}{N}\right)}.$$

Example (iii). Simple change of variable
We start with measurements of a and b which are uncorrelated, and which have variances $\sigma_a{}^2$ and $\sigma_b{}^2$ respectively. We want to calculate the error matrix on new variables x and y given by

$$\left.\begin{aligned} x &= \frac{1}{\sqrt{2}}(a+b), \\ y &= \frac{1}{\sqrt{2}}(a-b). \end{aligned}\right\} \tag{3.60}$$

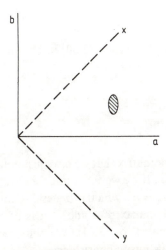

Fig. 3.11. A measurement has been made of two variables a and b, whose errors are uncorrelated. The shaded ellipse, whose axes are parallel to the a, b axes, indicates the range of likely values of a and b; σ_b is taken to be larger than σ_a. If we now choose x-, y-axes as shown (i.e. corresponding to the transformation (3.60) in the text), then the major axis of the ellipse is seen to have negative slope with respect to the new axes; if the value of x increases from its mean value (at the centre of the ellipse), the expected value of y decreases, i.e. $\text{cov}(x, y)$ is negative.

From eqn (3.46) we find that the new error matrix is given by

$$\begin{pmatrix} \dfrac{1}{\sqrt{2}} & \dfrac{1}{\sqrt{2}} \\ \dfrac{1}{\sqrt{2}} & -\dfrac{1}{\sqrt{2}} \end{pmatrix} \begin{pmatrix} \sigma_a^2 & 0 \\ 0 & \sigma_b^2 \end{pmatrix} \begin{pmatrix} \dfrac{1}{\sqrt{2}} & \dfrac{1}{\sqrt{2}} \\ \dfrac{1}{\sqrt{2}} & -\dfrac{1}{\sqrt{2}} \end{pmatrix}$$

$$= \frac{1}{2} \begin{pmatrix} \sigma_a^2 + \sigma_b^2 & \sigma_a^2 - \sigma_b^2 \\ \sigma_a^2 - \sigma_b^2 & \sigma_a^2 + \sigma_b^2 \end{pmatrix}. \quad (3.61)$$

Thus

$$\left. \begin{aligned} \sigma_x^2 &= \sigma_y^2 = \tfrac{1}{2}(\sigma_a^2 + \sigma_b^2) \\ \text{cov}(x, y) &= \tfrac{1}{2}(\sigma_a^2 - \sigma_b^2). \end{aligned} \right\} \quad (3.61')$$

and

This means that, unless σ_a and σ_b are equal, the errors on x and y are correlated. This can be appreciated from Fig. 3.11 where the transformation between the a, b and the x, y variables is shown. If σ_b is much larger than σ_a, the correlation is strong and *negative*; the shaded ellipse (corresponding to likely values of the variables) is orientated such that its axes are parallel to the a, b axes (i.e. no correlation) but its major axis has a *negative* gradient with respect to the x, y axes.

We can now invert the original transformation (3.60) to obtain

$$\left.\begin{aligned} a &= \frac{1}{\sqrt{2}}(x+y), \\ b &= \frac{1}{\sqrt{2}}(x-y), \end{aligned}\right\} \tag{3.62}$$

and use this, together with the x, y error matrix that we have just obtained, in order to deduce the a, b error matrix. If we were to omit the covariance term of (3.61), we would deduce that $\sigma_a = \sigma_b$, which in general is not true; in order to reobtain the correct error matrix of a and b, it is of course necessary to keep the whole error matrix of (3.61). We thus see that neglecting covariance terms can lead us into inconsistencies.

This example of transforming variables had already been dealt with to some extent in Section 1.5.1.

Example (iv). Another change of variables

If we measure the co-ordinates x and y of a point, we can transform to polar co-ordinates by

$$\left.\begin{aligned} r^2 &= x^2 + y^2, \\ \tan\theta &= y/x. \end{aligned}\right\} \tag{3.63}$$

If the errors on x and y are uncorrelated, we can calculate the r, θ error matrix from eqn (3.46) as

$$\begin{pmatrix} \sigma_r^2 & \mathrm{cov}\,(r,\theta) \\ \mathrm{cov}\,(r,\theta) & \sigma_\theta^2 \end{pmatrix} = \begin{pmatrix} \dfrac{x}{r} & \dfrac{y}{r} \\ -\dfrac{y}{r^2} & \dfrac{x}{r^2} \end{pmatrix} \begin{pmatrix} \sigma_x^2 & 0 \\ 0 & \sigma_y^2 \end{pmatrix} \begin{pmatrix} \dfrac{x}{r} & -\dfrac{y}{r^2} \\ \dfrac{y}{r} & \dfrac{x}{r^2} \end{pmatrix}$$

$$= \frac{1}{r^2} \begin{pmatrix} x^2\sigma_x^2 + y^2\sigma_y^2 & \dfrac{xy}{r}(\sigma_y^2 - \sigma_x^2) \\ \dfrac{xy}{r}(\sigma_y^2 - \sigma_x^2) & \dfrac{1}{r^2}(y^2\sigma_x^2 + x^2\sigma_y^2) \end{pmatrix}. \tag{3.64}$$

We note in passing that the transformation matrix is not symmetric, and hence we must get it the right way round;† and that the transformation

---

† If we accidentally interchange T and $\tilde{T}$ in eqn (3.46), then in this case the elements of the r, θ error matrix have incorrect dimensions, and hence the mistake should be obvious.

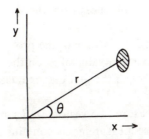

Fig. 3.12. A point in the first quadrant has been measured with the errors on the co-ordinates x and y uncorrelated, and with σ_y larger than σ_x. The shaded area shows the range of likely values of x and y. When transformed into polar co-ordinates, the shaded area has the property that as θ increases from its optimum value, the expectation value of r increases as well, i.e. the covariance of r and θ is positive in sign.

(3.63) is not linear and hence σ_x and σ_y must be small if the error matrix (3.64) is to be realistic.

We see from the r, θ error matrix in (3.64) that

$$\text{cov}\,(r, \theta) = \frac{xy}{r^3}\,(\sigma_y{}^2 - \sigma_x{}^2). \tag{3.65}$$

Thus for a point in the first quadrant, if σ_y is larger than σ_x, then the covariance is positive. That this is reasonable is seen from Fig. 3.12; the shaded area has the property that increasing θ corresponds to increasing r.

Problems

3.1 (i)   For values of r from zero to 20 inclusive, calculate the values of the probabilities $P(r)$ for a binomial distribution when the number of trials N is 50 and the probability of success is 0.2. Compare your answers with Fig. 3.1 (a). Estimate the width of the distribution, and compare it with the expected value.

(ii)   Calculate the corresponding probabilities for a Poisson distribution in which the mean number of successes is 10. Evaluate a Gaussian distribution of mean and variance both 10 and compare this with your Poisson distribution. What is the fractional area of the Gaussian distribution at negative values of r?

(iii) Repeat the procedure of part (i) for a binomial distribution

with $N = 20$ and $p = \frac{1}{2}$. Are the differences from the
distribution in (i) as expected?

3.2 Verify the statement in Section 3.4 that one of the properties of
the error matrix is that the maximum value of x' on the error
ellipse is given by the square root of the first diagonal element
of the error matrix.

3.3 In a certain experiment, we measure two vectors $(r_1 \, \theta_1)$ and
$(r_2 \, \theta_2)$. The error matrix of these measurements contains the
following terms:

$$\overline{\delta r_1{}^2} = \sigma_1{}^2 (4 \cos^2 \theta_1 + \sin^2 \theta_1),$$

$$\overline{\delta \theta_1{}^2} = \sigma_1{}^2 (4 \sin^2 \theta_1 + \cos^2 \theta_1)/r_1{}^2,$$

$$\overline{\delta r_1 \, \delta \theta_1} = -3\sigma_1{}^2 \sin \theta_1 \cos \theta_1 / r_1 = \overline{\delta \theta_1 \, \delta r_1},$$

$$\overline{\delta r_2{}^2} = \sigma_2{}^2 (4 \sin^2 \theta_2 + \cos^2 \theta_2),$$

$$\overline{\delta \theta_2{}^2} = \sigma_2{}^2 (4 \cos^2 \theta_2 + \sin^2 \theta_2)/r_2{}^2,$$

$$\overline{\delta r_2 \, \delta \theta_2} = 3\sigma_2{}^2 \sin \theta_2 \cos \theta_2 / r_2 = \overline{\delta \theta_2 \, \delta r_2}.$$

All other elements of the error matrix are zero.

(i)   Obtain the error matrix for the variables $x_1 \, y_1 \, x_2 \, y_2$ (i.e. where
the vectors are expressed in rectangular rather than polar
co-ordinates).

(ii) Calculate the error on the function $f = x_1 x_2 + y_1 y_2$.

(iii) Use the original error matrix for the variables $r_1 \, \theta_1 \, r_2 \, \theta_2$ to
obtain the error on the function $f = r_1 r_2 \cos(\theta_1 - \theta_2)$. Compare
your answer with that for part (ii).

3.4 In an experiment to measure the mass of a resonant state
whose decay position is known, the two decay particles (one
positively and one negatively charged) pass through a magnetic
field and then through some detectors, in order that their
momentum p and direction θ at the decay point can be
determined. For the positive track, the error matrix on the
variables $C_1(= 1/p)$ and θ_1 is

$$\begin{pmatrix} \delta C^2 & \delta C \, \delta \theta \\ \delta C \, \delta \theta & \delta \theta^2 \end{pmatrix},$$

where the elements are constants (i.e. independent of C_1 and
θ_1). For the negative track the error matrix is the same except
that the correlation term changes sign. (Why?)

The effective mass M of the decaying system is given by

$$M^2 = M_1^2 + M_2^2 + 2E_1 E_2 - 2p_1 p_2 \cos(\theta_1 - \theta_2),$$

where

$$E_i^2 = p_i^2 + M_i^2 \quad (i = 1 \text{ or } 2)$$

and the subscripts 1 and 2 refer to the positive and negative particles respectively, whose masses M_1 and M_2 are known. What is the error on M for a particular set of values p_1 θ_1 p_2 θ_2? Assuming that we are observing a resonance which decays symmetrically backwards and forwards, what is the influence of the correlation term $\delta C \, \delta \theta$ on the resolution after we average over all decay angles of the resonance?

3.5   An experiment is searching for quarks of charge $\frac{2}{3}$, which are expected to produce $\frac{4}{9}$ the ionisation I_0 of unit charged particles. In an exposure in which 10^5 cosmic rays are observed, 1 track has its ionisation measured as $0.44 I_0$. The apparatus is such that ionisation estimates are Gaussian distributed with standard deviation σ. Calculate the probability that this could be a fluctuation on the ionisation of a unit charged particle for the following different assumptions:

(a) $\sigma = 0.07 I_0$ for all 10^5 tracks,

(b)   for 99% of the tracks $\sigma = 0.07 I_0$, while the remainder have $\sigma = 0.14 I_0$.

4

Parameter fitting and hypothesis testing

As already mentioned in Chapter 2, the two basic sorts of problems that we deal with in the subject of statistics are hypothesis testing and parameter fitting. In the former, we test whether our data are consistent with a specific theory (which may contain some free parameters) and in the latter we use the data to determine the value(s) of the free parameter(s).

Thus consider that we have some data on an angular distribution, consisting of a set of values $\cos \theta_i$ for each interaction, where θ_i is the angle that the observed particle makes with some fixed direction. (We use $\cos \theta$ as variable rather than θ since an isotropic distribution in space corresponds to a flat distribution in $\cos \theta$, but not in θ.) Each $\cos \theta_i$-value could have associated with it a weight w_i greater than or equal to 1, to correct for detection inefficiencies. Then we can ask, for example, the following questions.

Question (i)
Are the data consistent with an angular distribution of the form

$$\frac{\mathrm{d}n}{\mathrm{d}\cos\theta} = a + b \cos^2 \theta? \tag{4.1}$$

And if the data look inconsistent with this, can we make a numerical estimate indicating how confident we are that the experimental data show that the angular distribution (4.1) is incorrect?

This is an example of hypothesis testing.

Question (ii)
Assuming that we accept the form of eqn (4.1) as reasonable for our experiment, what values of a and b should be chosen to provide the best description of our data? And how accurately are a and b determined?

This is an example of parameter fitting.

Logically, hypothesis testing precedes parameter fitting, since if our hypothesis is incorrect, then there is no point in determining the values of

74

the free parameters contained within the hypothesis. In fact, we will deal with parameter fitting first, since it is easier to understand. In practice, one often does parameter fitting first anyway; it may be impossible to perform a sensible test of the hypothesis before its free parameters have been set at their optimum values.

The main methods of parameter determination are those of moments, maximum likelihood and least squares, which we consider in Sections 4.3, 4.4 and 4.5 respectively.

4.1. Normalisation

In this section, we deal with the question of whether we wish to insist that the normalisation of the data and of the theory are the same. For the example of our angular distribution (4.1), this normalisation condition reduces to

$$N = \int_{-1}^{+1} (a + b \cos^2 \theta) \, d(\cos \theta)$$
$$= 2[a + b/3], \tag{4.2}$$

where N is the total number of observed events (weighted if necessary).

In deciding whether or not to normalise, some of the basic considerations are:

Consideration (i)
It is obviously desirable in many cases to normalise.

For example, if we rewrite our angular distribution as

$$\frac{dn}{d \cos \theta} = a \sin^2 \theta + (a + b) \cos^2 \theta \tag{4.1'}$$

it may describe the decay of a spin 1 nucleus, with the $\sin^2 \theta$ and $\cos^2 \theta$ terms respectively corresponding to the magnetic substates ± 1 and 0 respectively. Then $4a/3$ and $\frac{2}{3}(a + b)$ are the number of nuclei observed to be in their ± 1 and 0 substates respectively. Since this exhausts the possible substates for a spin 1 nucleus, we may want the sum of these numbers to equal the total number of decaying nuclei we have observed in our experiment.

Another example would be where we have fitted the fractions f_i of various subchannels contributing to a particular reaction. These fractions should obey the normalisation constraint that

$$\Sigma f_i = 1. \tag{4.3}$$

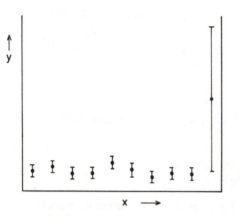

Fig. 4.1. A set of data points to which we wish to perform a straight line fit. In this case, it would be undesirable to insist that the theory and the data are normalised to each other, since the data point at the largest *x* value (and whose *y* co-ordinate has a very large error) distorts the data normalisation.

Consideration (*ii*)

The normalisation condition reduces the number of free parameters by one, and hence is useful. In our example, we started with two parameters (*a* and *b*), and can eliminate one of them immediately by the use of eqn (4.2).

Consideration (*iii*)

Some methods in some circumstances automatically normalise the theory to the data.

Consideration (*iv*)

In some cases, it is obviously undesirable. Thus, if we are trying to fit a straight line to the data of Fig. 4.1, then the point at the largest value of *x* contributes essentially no useful information as far as the straight line is concerned (because of its very large error). But in the normalisation condition, which involves calculating Σy_i for the data, each point is weighted equally; and so this can have a serious distorting effect. The data normalisation has a large error associated with it, but normalisation procedures do not take this into account. The best fit to the data of Fig. 4.1 in effect ignores the last point with the large error, and hence the normalisation of the fit is lower than that of the data in this case.

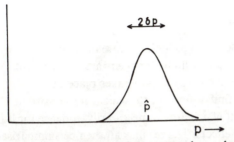

Fig. 4.2. A parameter, whose correct but unknown value is P_0, has been estimated as $\hat{p}$ with an uncertainty δp. The curve $y(p)$ is our confidence statement for that particular value of p being equal to the true value P_0. Provided that the data are such that the estimate $\hat{p}$ is Gaussian distributed, the range $\hat{p} \pm \delta p$ should contain P_0 in 68% of our experiments; and of all the ranges in p that have this probability, this is the shortest.

4.2 Interpretation of estimates

4.2.1 Meaning of error estimates

As a result of one of the methods to be described in the subsequent sections, let us assume that some parameter has been determined as $\hat{p} \pm \delta p$. We further assume that the data are such that our estimate of p is Gaussian distributed, and that the true fixed value of our parameter (which is probably unknown to us) is P_0.

For the actual value P_0 of the parameter, the probability that a measurement will give us an answer in a specific range of p (say from p_1 to p_2) is given by the area under the relevant part of the Gaussian curve. A conventional choice for this probability is 68%. Of all the ranges which possess this property, the one defined by

$$P_0 - \delta p \leqslant p \leqslant P_0 + \delta p \tag{4.4}$$

is the shortest in p (assuming a Gaussian distribution), and hence the most useful in general.

Having obtained an estimate $\hat{p}$, it is usual to rewrite (4.4)† as

$$\hat{p} - \delta p \leqslant P_0 \leqslant \hat{p} + \delta p \tag{4.5}$$

and then to think of this as a confidence range for P_0, i.e. how often we expect to include P_0 within our quoted range, for a repeated series of experiments. In principle, however, (4.5) is still a probability statement about the random variable $\hat{p}$, rather than on the fixed parameter P_0, which simply is or is not within any specific range.

† The relationship of (4.5) to (4.4) is that if $\hat{p}$ is close to P_0, then P_0 is close to $\hat{p}$.

4.2.2 Upper limits

Some experiments are designed to look for rare or 'forbidden' processes, and conclude by quoting upper limits for their rate. For example, we may be interested in whether the decay $\mu \to e\gamma$ takes place or not. In general, such experiments may find a few events which are consistent with the searched-for process, but which are not necessarily evidence for it because of possible background effects. The results can then be summarised as an estimated branching ratio for the process which could well be of the form $(3 \pm 5)\ 10^{-9}$, i.e. consistent with zero.

If it is assumed that the experiment is such that the distribution of the measured rate is Gaussian, then from Fig. 3.4, we can deduce that there is a 90% probability of the measured value B_m being not more than 1.28 standard deviations (σ) below the true value B_0. Thus, at the 90% confidence level

$$B_0 < B_m + 1.28\sigma. \tag{4.6}$$

The corresponding factors for 95% and 99% confidence are 1.64 and 2.33 respectively. Such formulae are often used to convert the observed rate into an estimated upper limit on the actual rate (see Fig. 4.3(a)).

In fact the actual estimated rate and its error are more useful than simply the upper limit. Thus if we wish to combine the results of different experiments, the upper limits on their own cannot be used, while the estimates, even if consistent with zero, can be averaged according to formulae (1.38) and (1.39).

There is furthermore a problem of how to allow for the fact that a significant part of the range defined by such a procedure for B_0 may be unphysical. Thus in our example, B corresponded to a branching ratio which is necessarily positive. This effect is even more serious when the estimated B_m turns out to be negative, which it can do experimentally because of statistical fluctuations associated with background subtraction procedures. The approach of simply adding 1.28σ to the negative B_m (see Fig. 4.3(b)) can result in an artificially low estimated upper limit for B_0.†
Indeed if B_m is sufficiently negative, our 90% or 95% confidence limit for B_0 could itself even be negative.

---

† The logic of this approach is that for some specific small value of B_0, the probability of obtaining a significantly negative value of B_m is much smaller than for a measured value B_m close to B_0. Since, however, any positive value of B_0 would result in a small probability of observing a significantly negative value of B_m, the usefulness of a negative B_m in discriminating between different possible values of B_0 may well be exaggerated by this procedure.

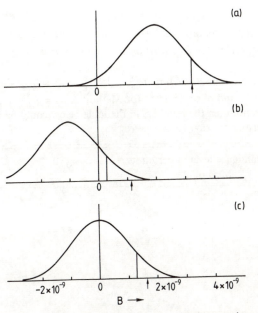

Fig. 4.3. Gaussian distributions centred on various measured values B_m of a branching ratio (a) $B_m = (2 \pm 1)\ 10^{-9}$, (b) $B_m = (-1 \pm 1)\ 10^{-9}$, (c) $B_m = (0 \pm 1)\ 10^{-9}$. The solid vertical lines on the Gaussians indicate the 90% confidence upper limits on the actual value of the branching ratio, as deduced from formula (4.6). The arrows show where the area under the Gaussian from zero up to that point is equal to 90% of the area for $B > 0$. Whereas these two estimates are essentially identical in (a), they differ significantly in (b) and are slightly separated even in (c).

A more sensible method would seem to be to set a 90% confidence level upper limit on B_0 at a value such that

$$A_1/A_0 = 0.9,$$

where A_1 is the area under a Gaussian (centred on B_m and of width σ) within the physical region† for B and up to our suggested 90% confidence estimate, and A_0 is the corresponding area within the whole physical region‡ (see Fig. 4.3). Such a procedure would also be relevant even for B_m being zero or slightly positive (see Fig. 4.3(c)).

---

† In some situations, it may also be necessary to take into account a possible upper limit on the physical range of B, e.g. $0 \leqslant B \leqslant 1$.

‡ Our approach is similar in spirit to calculating the likelihood (as a function of the branching ratio) for obtaining the observed number of events N_{obs}, when the calculated background N_b is larger than N_{obs}.

Table 4.1. *Comparison of two different methods for deducing a* 90%
confidence limit for a parameter which is necessarily positive, as a
function of the measured value B_m, *with an assumed error* σ *of* ± 1

Method 1 determines the limit B_ℓ such that $\int_0^{B_\ell} g\,dB / \int_0^{\infty} g\,dB = 0.9$, where g is a
Gaussian centred on B_m and of unit width; for Method 2, $B_\ell = B_m + 1.28\sigma$. If
B were a branching ratio, then the numbers in the table below could be taken
as all needing multiplying by, say, 10^{-9}.

| B_m | 90% confidence level upper limits | |
	Method 1	Method 2
5	6.3	6.3
4	5.3	5.3
3	4.3	4.3
2	3.3	3.3
1	2.4	2.3
0.5	2.0	1.8
0	1.6	1.3
-0.5	1.4	0.8
-1	1.2	0.3
-2	0.8	(-0.7)
-3	0.6	(-1.7)
-4	0.5	(-2.7)
-5	0.5	(-3.7)

In Table 4.1, we show the difference in using these two procedures for
estimating upper limits.

Exactly analogous remarks of course apply to estimates of lower limits.

4.2.3 *Estimates outside the physical range*

A related question to that discussed above is whether we are prepared to
accept an estimated value *outside* the allowed physical range for the
parameter in question, e.g. a fraction of some population is estimated as
larger than unity or less than zero. In some estimation procedures (e.g.
maximum likelihood – see Section 4.4), the best estimate is sometimes
required to be within the physical range, while in others (e.g. the method
of moments – see Section 4.3) this is not so.

Clearly if our estimated fraction is 1.1, then the most sensible estimate
as far as our data are concerned is 1.0. If, however, we are interested in

combining the results of different experiments measuring the same quantity, it can be a mistake to force all estimates above 1 to be unity while leaving alone all those below 1. This would produce a biased average in the case where the true value of the fraction were close to unity, and the estimates above and below 1 arose simply from statistical fluctuations on the finite data samples. If the experiments were combined at the data sample levels (rather than simply by combining their estimates), this problem of course does not arise.

4.3. The method of moments

The method of moments consists in calculating the average value for the data of some suitably chosen quantity. This average is analytically related to, and hence allows the determination of, the parameter of interest.

For our angular distribution (4.1), we find by integration that

$$\overline{\cos^2 \theta} = \frac{5 + 3b/a}{5(3 + b/a)},\qquad(4.7)$$

whence

$$b/a = \frac{5(3\,\overline{\cos^2 \theta} - 1)}{3 - 5\,\overline{\cos^2 \theta}}.\qquad(4.7')$$

(It is always useful to perform simple tests where possible to check the algebra. In this case we note that, for an isotropic angular distribution, $\overline{\cos^2 \theta}$ is $\frac{1}{3}$ and b/a is zero. These are consistent with (4.7) and (4.7').)

So in order to evaluate b/a from our data, we calculate

$$\overline{\cos^2 \theta} = \frac{1}{n} \sum_{i=1}^{n} \cos^2 \theta,\qquad(4.8)$$

and the error δ on this average

$$\delta = \frac{1}{\sqrt{n}} \sqrt{\left[\frac{1}{n-1} \sum_{i=1}^{n} (\cos^2 \theta_i - \overline{\cos^2 \theta})^2 \right]},\qquad(4.9)$$

(Compare Section 1.4.2.) In conjunction with (4.7'), this then gives b/a and its error.

In order to calculate b and a separately rather than just their ratio, we can use the normalisation condition (4.2).

Comment (i). Simplicity
The method of moments is very easy to apply, and it does not require a maximisation procedure. If we were to perform such a calculation only

once, this would be only a slight advantage, but in real life situations we may wish to determine the value of b/a as a function of the mass of the system of interest and of its production angle, etc., and subject to various other cuts. Furthermore, we may want to fit functional forms other than just (4.1); these could, for example, involve the azimuthal angle ϕ as well as the polar angle θ. So we can easily be involved in calculating the values of about 100 different parameters, even in fairly simple situations. By the method of moments this is fairly trivial, but to obtain them via techniques that require maximisation is likely to be a major operation.

On the other hand, the method of moments does rely on the fact that we can calculate a suitable moment of the theoretical distribution in order to obtain the required parameter. In complicated problems, the required integration may turn out to be impossible to do. Thus the moments method has its greatest application in problems involving relatively simple distributions.

Comment (ii). Error estimates

The error is easily obtained from the spread on the observed values of $\cos^2 \theta_i$; or alternatively, from the fitted distribution, we can calculate the expected spread as a function of b/a. In the first method, the spread can be calculated at the same time we are looping over the events in order to calculate the mean. But it suffers from the disadvantage that it can accidentally give a ridiculously small error (especially if the fitted distribution is incorrect). For example, if all the $\cos \theta_i$ are equal, the spread in values and hence also the error estimate are both zero. The second method does involve more work, especially if the necessary integration required to estimate the spread is difficult.

In either of these methods, it is necessary to remember to divide the observed or calculated spread of the distribution by $\sqrt{n}$, where n is the observed number of events. If the events have varying weights arising from detection inefficiencies, then instead we divide by $\sqrt{n_{eff}}$, where n_{eff} is the effective number of events as defined in equation (1.13).

The error is symmetric in the variable $\overline{\cos^2 \theta}$. We must of course transform to the variable b/a by the non-linear relation (4.7'). Thus the error on b/a will in general be asymmetric. This of course is physically reasonable; the distribution for $b/a = 50$ is very different from that for $b/a = 0$, but rather similar to that for $b/a = 100$.

Comment (iii). Individual events

The moments procedure makes use of individual events, and hence there is no need to make a possibly compromising choice of bin size with which to plot a histogram.

Comment (iv). Non-uniqueness

A problem associated with the method is that in complex situations, there are often many possible distributions of which we can take moments, which can then over-determine the parameters and provide inconsistent estimates of them. Even for a single distribution like (4.1), we can calculate all the moments $\cos^{2K}$ (where K is any positive integer), each of which provides a linear relation between a and b. A tedious calculation, however, shows that the intuitive choice of $2K = 2$ gives an estimate of b/a with an error which is generally smaller than for any other choice of K.

Comment (v). Unphysical answers

As can be seen from eqn (4.7), if

$$\overline{\cos^2 \theta} \leqslant \tfrac{1}{5}, \tag{4.10}$$

then

$$b/a \leqslant -1 \tag{4.10'}$$

and the angular distribution (4.1) will become negative for part of the $\cos \theta$ range. This is clearly unphysical, but from statistical fluctuations or because the distribution is not of the form (4.1) it is possible that a given sample of events will violate the bound (4.10). Similarly, as

$$b/a \to \infty, \tag{4.11}$$

$$\overline{\cos^2 \theta} \to \tfrac{3}{5}, \tag{4.11'}$$

and any larger value of $\overline{\cos^2 \theta}$ is unphysical (and yields very negative b/a).

The above are fairly obvious examples of physical bounds. A slightly more subtle case arises in the parity violating decay of a spin $\tfrac{1}{2}$ particle; the decay distribution is given by

$$\frac{\mathrm{d}n}{\mathrm{d} \cos \theta} = N(1 + \alpha P \cos \theta), \tag{4.12}$$

where N is a normalisation constant, P is the polarisation of the particle in the given reaction, and α is a number intrinsic to the particle and whose modulus is smaller than unity. Since the angular distribution (4.12) must stay positive for $|\cos\theta|$ less than unity, we have a mathematical constraint that

$$|\alpha P| \leqslant 1, \tag{4.13}$$

which implies that

$$|\overline{\cos\theta}| \leqslant \tfrac{1}{3}. \tag{4.13'}$$

But for particles for which

$$|\alpha| < 1, \tag{4.14}$$

even for moments satisfying (4.13'), we can still obtain

$$|P| > 1, \tag{4.14'}$$

which is physically impossible. The moment must in fact satisfy the stronger constraint

$$|\overline{\cos\theta}| < \frac{|\alpha|}{3}. \tag{4.14''}$$

In other cases, constraints arise among different parameters. Thus for example, if our angular distribution has the form

$$\frac{\mathrm{d}n}{\mathrm{d}\cos\theta} = N(1 + \alpha\cos\theta + \beta\cos^2\theta),$$

rather than (4.1), then there are constraints on α and β to ensure that the distribution remains positive within the physical range. But α and β are determined separately from the moments $\overline{\cos\theta}$ and $\overline{\cos^2\theta}$, and hence can in practice violate the constraints.

In general, any form of constraint on the parameter or parameters is difficult to include in the moments method.

Comment (vi). Inefficiencies

If our apparatus is such that our data does not extend over all values of $\cos\theta$, we can evaluate $\overline{\cos^2\theta}$ over a more restricted range (e.g. for $\cos\theta$ between -0.8 and $+0.8$), and then obtain our parameters b and a from appropriately modified versions of eqns (4.2) and (4.7).

If we have a variable detection efficiency $\varepsilon(\cos\theta)$, we can use weights $w(=1/\varepsilon)$ and calculate a weighted average for $\cos^2\theta$ (see Section 1.4.2). This should correct for the losses caused by apparatus inefficiences.

Comment (vii). Hypothesis testing

In the moments method, there is no direct check that can be made as to whether the distribution whose parameters are being determined is reasonable or not.

Thus in summary the moments method is very useful as a fast and simple method for obtaining some idea of the values of parameters, which may however be slightly unphysical in their magnitudes. One use of the moments method is in deducing starting values of the parameters, to be used in more complicated fitting procedures, thereby reducing the computer time spent in the latter method.

4.4. Maximum likelihood method

4.4.1 What is it? Some simple examples

The maximum likelihood method is the most powerful one for finding the values of unknown parameters. In order to understand how it works, we return to the specific example of the angular distribution of eqn (4.1).

We begin by rewriting the angular distribution as

$$y = \frac{\mathrm{d}n}{\mathrm{d}\cos\theta} = N(1 + (b/a)\cos^2\theta), \tag{4.15}$$

where N is a normalisation factor such that

$$\int_{-1}^{+1} y \, \mathrm{d}\cos\theta = 1, \tag{4.16}$$

i.e.

$$N = \frac{1}{2(1 + (b/3a))}. \tag{4.17}$$

Thus we have normalised y so that it behaves as a probability distribution. We stress that it is *essential* to remember to include the normalisation factor N in (4.15) (at least as far as its dependence on the parameter b/a is concerned – the constant factor of $\frac{1}{2}$ in eqn (4.17) is not crucial) or else the maximum likelihood method will not work. We wish to determine the one parameter b/a in eqn (4.15).

For the ith event we calculate

$$y_i = N(1 + (b/a)\cos^2\theta_i), \tag{4.18}$$

which is the probability density for observing that event and is a function of b/a. Then we define the likelihood $\mathcal{L}$ as the product of the y_i for all the events of our sample i.e.

$$\mathscr{L}\left(\frac{b}{a}\right) = \prod_{i=1}^{n} y_i, \tag{4.19}$$

with the y_i as defined in eqn (4.18). Thus, for any specific value of b/a, $\mathscr{L}$ is the joint probability density for obtaining the particular set of $\cos \theta_i$ values that we have observed in our experiment.† So we finally maximise $\mathscr{L}$ as a function of b/a in order to find the best value of b/a. We now see why the normalisation is essential; without the factor of N in eqn (4.15), we can make all the y_i larger simply by increasing b/a, and hence the function $\mathscr{L}$ would have no absolute maximum.

As a second example of the maximum likelihood method, we consider the case of fitting a Breit–Wigner resonance curve to a set of mass values m_i. We write the Breit–Wigner as‡

$$y_i(M_0, \Gamma) = \frac{1}{2\pi} \frac{\Gamma}{(m_i - M_0)^2 + (\Gamma/2)^2}. \tag{4.20}$$

We assume that Γ is known and that there is no background present, so that the only parameter to be determined is M_0. We can then suppress the Γ dependence of y, and define the likelihood for this example as

$$\mathscr{L}(M_0) = \prod_{i=1}^{n} y_i(M_0). \tag{4.21}$$

The maximum of $\mathscr{L}$ gives the best value of M_0.

We can understand how the method works by looking at Fig. 4.4(a). The observed mass values are indicated by the series of short bars along the mass axis. The curve is a plot of eqn (4.20) for a particular M_0-value. As we vary M_0, we simply shift the whole curve sideways without changing its shape or height. (The value of Γ is fixed, and the normalisation is independent of the choice of M_0.) Then the likelihood method consists in finding a position of the curve such that the product of the y_i-values at

---

† Eqn (4.19) as it stands is the probability density for obtaining our particular set of observations in the order in which we observed them. Since the order is in fact irrelevant, there should be a factor of $n!$ on the right-hand side of the equation. Since, however, we are interested only in how $\mathscr{L}$ varies with the parameter(s) we are determining, a constant multiplicative factor is irrelevant.

‡ Formula (4.20) is already normalised, i.e. its integral over m is unity. If we regard y as a function of M_0 only (i.e. Γ fixed) then the factor of $\Gamma/2\pi$ is constant and can be omitted. But if we wish to determine Γ, then the factor of Γ in the numerator is essential.

The y_i of formula (4.20) is clearly also a function of the experimental mass m_i but we omit explicit mention of this in the arguments of y_i since we wish to emphasise the dependence on the parameters M_0 and Γ.

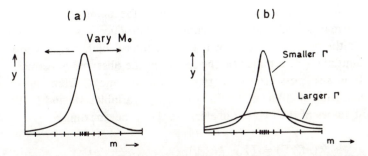

Fig. 4.4. A visual representation to show how the maximum likelihood method succeeds in finding the best value of the parameters. The bars along the m-axis represent experimental measurements of a set of mass values m_i, which are fitted by a Breit–Wigner resonance shape. In (a) the width of the resonance is kept fixed and the central mass M_0 is varied. This has the effect of sliding the $y(M_0, \Gamma)$ curve along the m-axis, while keeping its height and shape unaltered. For a given position of the curve, we multiply together all the y_i values (i.e. y evaluated at each of the m_i) to obtain the likelihood function $\mathscr{L}$. The value of M_0 which maximises $\mathscr{L}$ is then the best estimate of the mass of the state. This clearly occurs when the curve is centred on the concentration of experimental mass values. In (b), the mass of the state is kept constant but its width is varied. Since the curves are normalised to constant area, the curve for smaller Γ is higher at the central value than is that for larger Γ. The best value of Γ is that for which the product of the y_i-values is largest. In this case it is the spread in the experimental mass values which is effective in determining Γ.

the positions of the bars is a maximum. Clearly, we will get the largest value if the curve is high where most of the bars are concentrated. Thus the best value of M_0 is in the region where the largest number of masses are found.

How does the method work when we want to find the width Γ? We now fix M_0, and vary Γ. Thus we here regard (4.20) as displaying the Γ dependence of y_i, and maximise

$$\mathscr{L}(\Gamma) = \prod_{i=1}^{n} y_i(\Gamma). \tag{4.21'}$$

In Fig. 4.4(b), the effect of varying Γ is demonstrated. Not only does the shape of the curve change, but so does its height at the maximum (in order to preserve the normalisation condition). The need to have the curve large where most of the events are concentrated means that Γ must not be too large (since then the y_i for the main concentration of events are all low), nor must it be too small (since although a few events near the centre of the observed distribution will have large y_i, the majority of the events will have small y_i and since $\mathscr{L}$ is defined as a product, it will be very small). Thus the optimum choice of Γ is some intermediate value whose exact magnitude is determined by the spread of the observed m_i-values.

It is also possible to determine both M_0 and Γ from the data by considering (4.20) as a function of the two variables M_0 and Γ. Then the best values of the two parameters are those that maximise

$$\mathscr{L}(M_0, \Gamma) = \prod_{i-1}^{n} y_i(M_0, \Gamma). \tag{4.21''}$$

4.4.2 The logarithm of the likelihood function, and error estimates

In most problems, it is usually convenient to consider the logarithm of the likelihood function

$$\ell = \log \mathscr{L} \tag{4.22}$$

$$= \sum_{i-1}^{n} \log y_i. \tag{4.22'}$$

For a large number of experimental observations n, $\mathscr{L}$ may tend to a Gaussian distribution (at least in the vicinity of the maximum of the distribution). A not very convincing proof of this is provided by performing a Taylor series expansion of ℓ near its maximum. Then

$$\ell = \ell_{max} + \frac{1}{2!} \ell'' \delta p^2 + \dots \tag{4.23}$$

$$= \ell_{max} - \frac{1}{2c} \delta p^2 + \dots, \tag{4.23'}$$

where in the last step we have simply replaced the second derivative ℓ'' by $-1/c$. Then from (4.22)

$$\mathscr{L} \sim \exp\left(-\frac{(p - p_0)^2}{2c}\right), \tag{4.24}$$

as we set out to demonstrate. The relation between ℓ and $\mathscr{L}$ in such a case is shown in Fig. 4.5.

The best value of p is that which maximises $\mathscr{L}$ (or ℓ). The range of likely values of the parameter p is determined by the width of the $\mathscr{L}$ distribution. If this is sharply peaked around its maximum, then so are the probable p values, since p values further away correspond to a very low probability

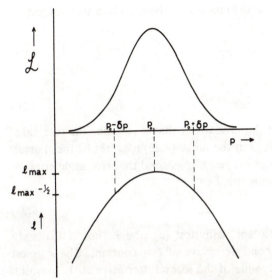

Fig. 4.5. In many situations, ℓ (the logarithm of the likelihood function) tends to be parabolic as a function of the parameter p in the region of the maximum. In that case, the likelihood function $\mathscr{L}$ is Gaussian distributed. The best estimate of the parameter p_0 is that which maximises either $\mathscr{L}$ or ℓ. The accuracy δp with which p_0 is determined is defined by the condition $\ell(p_0 \pm \delta p) = \ell(p_0) - \frac{1}{2}$. This is here equivalent to the statement that δp^2 is the variance of the $\mathscr{L}$ distribution.

for obtaining our particular set of experimental observations. In cases where the $\mathscr{L}$ distribution is Gaussian, the following quantities are identical, and any of them can be used as the definition of the error on p:

(i) the root mean square deviation of the $\mathscr{L}$ distribution about its mean;

(ii) $[(-\partial^2\ell)/(\partial p^2)]^{-\frac{1}{2}}$, i.e. the reciprocal of the square root of the negative of the second derivative of the ℓ distribution; or

(iii) the change in p required to reduce ℓ from its maximum value by 0.5, i.e.

$$\ell(p_0 \pm \delta p) = \ell(p_0) - \frac{1}{2}.$$

(N.B. This is not a factor of 0.5, but simply a difference of 0.5.)

What do we do in cases where the distribution of $\mathscr{L}$ in p is not Gaussian?

One possibility is to use the definition (iii) above to find two values† p_1 and p_2 such that

$$\ell(p_i) = \ell(p_0) - \tfrac{1}{2}, \quad i = 1 \text{ or } 2. \tag{4.25}$$

Then the range

$$p_1 \leqslant p \leqslant p_2 \tag{4.26}$$

still has the property that the probability that it contains the true value P of the parameter is 68%; but the range (4.26) may not be the shortest one in p which possesses this property. In general the error as determined in this way will not be symmetric, i.e.

$$p_0 \neq (p_1 + p_2)/2. \tag{4.27}$$

Alternatively, we can adapt definition (ii) above. For a Gaussian distribution of $\mathscr{L}$, the second derivative of ℓ is constant. The simplest modification is to use the value of the second derivative at the position of the maximum. It is probably more sensible to use

$$\left\langle \left(-\frac{\partial^2 \ell}{\partial p^2} \right)^{-\frac{1}{2}} \right\rangle,$$

where the average extends over a suitable range of p, e.g. over that range of p for which $\mathscr{L}$ is large. A crude justification for this is as follows. If $\mathscr{L}$ falls off more slowly than a Gaussian distribution, then the error is larger and also $\langle (-\partial^2 \ell)/(\partial p^2) \rangle$ is smaller away from the maximum; so the procedure seems sensible. This clearly is going to be messier to perform than simply to base our estimate of the value of the second derivative at the maximum.

It is clear that non-Gaussian situations are unpleasant, and so if possible it is highly desirable to choose our variables sensibly (i.e. to make the error distribution more like a Gaussian). Thus, for example, in measuring decay processes, it may well be better to calculate the decay rate $(1/\tau)$ rather than the lifetime τ itself. Another example is provided by determining the momenta p of charged tracks by measuring how much they are bent in a magnetic field. Then the errors look much more sensible if we calculate $1/p$ rather than p.

† If the shape of the ℓ distribution is complicated, there may be more than two values which satisfy (4.25).

4.4.3 *Comments on the maximum likelihood method*

Comment (i)

The maximum likelihood method uses the events one at a time, and so there is no need first to construct a histogram (with its associated problem of what bin size to choose). It also means that the method can be used when the density of events over the physical region is low, and for methods that require histograms, there would not be sufficient events per bin to make these alternative methods usable.

Comment (ii)

In many circumstances the maximum likelihood approach is the most efficient method of analysing one's data. Again, this is an especially useful feature for dealing with experiments involving only a small number of events and where it is essential to extract the maximum possible information from them.

Comment (iii)

We can transform from one variable to another, and the answer remains unique. Thus if we determine a decay rate as $\lambda_0 \pm \delta\lambda$, or if we use the same data to extract the lifetime as $\tau_0 \pm \delta\tau$, then

$$\lambda_0 = \frac{1}{\tau_0}. \tag{4.28}$$

If we define the errors by the '$\ell_{max} - \frac{1}{2}$' method, then the errors correspond as well, i.e.

$$\lambda_0 + \delta\lambda = \frac{1}{\tau_0 - \delta\tau} \tag{4.29}$$

(although this is not necessarily so if we alternatively use the value of the second derivative of the logarithm of the likelihood function to determine the errors).

Comment (iv)

Functions of implicit variables are very easily handled. All that is required is an extra line of computation. In contrast, the moments method becomes very difficult to apply for any but the simplest functional forms.

Comment (v)

The maximum likelihood method is a very powerful tool for determining unknown parameters in physical theories, as the data are used in the form of complete events rather than as projections on various axes. Thus, for example:

(a) A decay distribution is known to be of the form $1+3\cos^2\theta$ with respect to a certain polar axis whose direction we wish to determine. In the maximum likelihood method, the axis is defined by two independent variables which we proceed to determine. In other methods, we must use projections of the experimental data onto various arbitrary sets of axes (and how many of these should we use?).

(b) A given physical process is supposedly described by a certain theory, from which we can obtain a matrix element, which in turn contains some arbitrary parameters. We wish to use our data to determine these parameters. By the likelihood method, this is a straightforward if somewhat tiresome task with each event in turn being considered as a whole. In other methods, we need to compare the experimental *distributions* in some kinematic variables with the predictions of the theory. Once again we are faced with the problem of which variables to use for our histograms, and how many of them are worth comparing. (Compare Note (v) on pages 108/9.)

Comment (vi)

We frequently need to fit some specified functional form to a set of data points displayed in the x–y plane, whose errors in the y-direction are Gaussian distributed but whose x-values are precisely determined. In this situation the maximum likelihood and the least squares methods are equivalent.

Comment (vii)

If any of the parameters are constrained to be within a certain physical range (for example, greater than zero, or between ± 1, etc.) or if there are constraints among the parameters (for example, the sum of their squares is unity), then in the maximum likelihood method these are generally easily imposed; we simply do not allow the parameters to move outside the allowed domain while we are searching for the maximum. Problems can arise, however. Thus if the function is largest on the edge of the physical region, then it is not a maximum there in the strict mathematical sense (i.e. the first derivative is not zero). In general, it can be troublesome to find the maximum if it is situated near to, on, or outside the boundary of the physical region. (When a parameter does become unphysical, ℓ may become equal to the logarithm of a negative number.)

Comment (viii)

In the likelihood method, background subtraction techniques† can be a nuisance. We are faced with a dilemma as to how to proceed.

(a) We could calculate the values of the parameters first for our signal region, and then for the background control region; and use a subtraction method to extract the best values of the parameters to describe our background-free signal. But there is no particular reason why the same functional form that describes our signal should be applicable to the background region, and so its use there (as required in the second stage described above) is of dubious validity.

(b) We could calculate

$$\ell' = \ell_S - \ell_B,$$

where ℓ_S is obtained by summing over events in the signal region and ℓ_B is defined similarly for events in the background region. The practical problem that this can present is as follows. For specific values of the parameters, a particular event can have a set of observed variables such that it results in a very small value of y_i, and hence a very negative value of the corresponding ℓ. If this event is in the background region, it produces a large positive contribution to ℓ', and hence can distort or destroy the maximum that we are seeking.

The best procedure is not to perform a background subtraction as such but to define a more complicated likelihood function. Thus in the familiar case of a resonance sitting on top of some background, we are interested in determining some properties associated with the resonance (for example, some parameters relating to the distribution in the angle at which the resonance is produced, i.e. between the incident beam particle and the outgoing resonance). Then we would define a likelihood function

$$\ell = \sum_i \log\{f_R w_R(m_i) p_R(\theta_i) + f_B w_B(m_i) p_B(\theta_i)\}, \qquad (4.30)$$

where

f_R is proportional to the fraction of resonant events,
m_i is the observed value of the mass for the ith event,
w_R is the functional form used to describe the resonance,
θ_i is the observed value of the production angle,

† Background subtraction is discussed in more detail in Appendix 1.

p_R is the functional form used to describe the resonance's production angular distribution, and involves the parameters to be determined,

and

f_B, w_B and p_B are the corresponding parameters or functions for the background.

We have thus overcome the previous deficiencies, but at the expense of much greater complexity. Thus we have to choose adequate functional forms for w_B and for p_B (as well as for w_R and p_R). We then have a larger number of parameters, with respect to which our likelihood function must be maximised.

Comment (ix)

In some cases, the experimental data consist of events each with its own weighting factor w_i allowing for a variable detection efficiency of our apparatus (i.e. $w_i \geq 1$). Finding the best values of the parameters is straightforward even in this case: we can define

$$\ell = \sum_i w_i \log(y_i) \tag{4.31}$$

as giving the likelihood function for the observed data, corrected to what it would have looked like if the apparatus had been fully efficient.

The problem arises in finding the error on the parameters. Straightforward application of the standard error recipe produces error estimates that are *smaller* than those which would be obtained if all the w_i were set equal to unity. This is clearly incorrect, since apparatus *inefficiences* cannot produce *improved* estimates of parameters.

It is thus preferable to determine the parameters by constructing the likelihood function for the observed uncorrected data as follows:

$$\ell_b = \sum_i \log(N y_i \varepsilon_i), \tag{4.31'}$$

where $\varepsilon_i = 1/w_i$ is the efficiency for observing the ith event, $y\varepsilon$ gives the expected distribution allowing for the apparatus inefficiencies and N is the normalisation factor for the function $y\varepsilon$. Then the parameters and their errors can be determined from ℓ_b in the usual way.

The difference of these two approaches is brought out by problem 5.3.

Comment (x)

Whereas we can calculate the mean value of any experimental quantity in a model-independent way, the likelihood method (and the least squares one) require an assumed functional form before anything can be deduced. Thus if we have an experimental histogram of the number of observed interactions for a specified beam and target as a function of the number of charged tracks (i.e. a multiplicity distribution), we can average and obtain the mean multiplicity and its error uniquely, whereas the results of the likelihood and least squares methods may depend on whether the distribution is assumed to be Gaussian, Poisson, Poisson in the number of negative charged tracks, etc.

A compensating feature occurs when our data exhibit overflow, i.e. when for some events the variable is measured simply as being larger than a specified value. This can occur, for example, if an analogue signal is being digitised, but its amplitude falls outside the range set for the analogue-to-digital conversion. Then our data will consist of a set of actual values, plus the number of overflow measurements. It is now impossible to calculate the mean directly. The moment method copes with this by assuming a specific functional form; it ignores the overflow, and a suitable correction is applied. In the likelihood and least squares methods, however, the overflow events are actually included in the fit; even though their exact values are unknown, they are compared with the theoretical prediction, integrated over the relevant range of the overflow bin.

Comment (xi)

One of the most serious drawbacks of the likelihood method is that in general it requires a large amount of computation. This arises because the method involves a maximisation, often as a function of several parameters. Another aspect that can require a lot of calculation is the overall normalisation factor which enters into the definition of the likelihood; in all but the simplest problems, this factor cannot be obtained analytically, and hence must be estimated numerically. Since the normalisation factor is in general a function of the parameters (otherwise it is just a constant, and can be omitted), it has to be re-evaluated at each stage of the maximisation as the parameters change their values.

Comment (xii)

Another drawback is that it is in general difficult to know whether the functional form used to fit the data is satisfactory or not, i.e. does the function (with the best estimate of the parameters) provide a good fit to the data? If the fit is 'good', then the value of ℓ at its maximum is 'large', while a 'poor' fit produces a 'small' value of ℓ_{max}. Clearly the problem is to decide 'How large is large?'. For very simple problems, this can be answered analytically. A method involving less thought (but much more computer time) is to perform a Monte Carlo† series of calculations after the real data has been fitted. We randomly generate a series of artificial data sets according to the known functional form with the fitted values of the parameters, use the likelihood method to determine ℓ_{max} for each of them, then plot the distribution of these ℓ_{max}-values; then we can see how the value of ℓ_{max} from the actual data compares in magnitude with those that correspond to samples from a known distribution.

In order to obtain a reasonable distribution of ℓ_{max}-values we require many (i.e. 10–1000, depending on how the actual ℓ_{max}-value compares with the Monte Carlo ones, and at what level we wish to make our statistical statement) sets of Monte Carlo data. Thus, as well as all the work of generating these samples, we have to perform our maximum likelihood search the corresponding number of times. (Admittedly we know the approximate locations of the maxima for these generated sets.) Clearly we are going to be involved in large amounts of computing.

In most cases, the hypothesis testing question is more simply (albeit somewhat less satisfactorily) answered by a method based on a χ^2 technique (see Section 4.6).

4.4.4 Several parameters

When ℓ was a function of one parameter p, the best estimate of the parameter was determined by

$$\frac{d\ell}{dp} = 0$$

and, at least for the case where ℓ was parabolic near the maximum (i.e. $\mathscr{L}$ was Gaussian-shaped), the error

$$\sigma = \left(-\frac{d^2\ell}{dp^2} \right)^{-\frac{1}{2}}.$$

When ℓ is a function of several variables p_i, their best values are determined by the simultaneous equations

$$\frac{\partial \ell}{\partial p_i} = 0. \tag{4.32}$$

As far as the errors are concerned, we define

$$H_{ij} = -\frac{\partial^2 \ell}{\partial p_i \, \partial p_j}. \tag{4.33}$$

† See Chapter 6 for more details of the Monte Carlo technique.

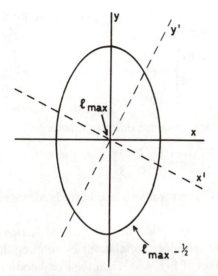

Fig. 4.6. A log-likelihood function ℓ of two variables. Here ℓ maximises at the origin. The ellipse shown corresponds to the contour for $\ell_{max} - \frac{1}{2}$, which is used to obtain the errors on the variables x and y. Since the axes of the ellipse are parallel to the x and y axes, the errors are uncorrelated. The simplest way to produce the effect of correlations is to rotate the axes. Then ℓ still maximises at the origin, but we see that the ellipse is now inclined. As y' increases from its optimum value (zero), the value of x' required to maximise ℓ decreases; this is a negative correlation. The magnitudes of the errors (including their possible correlation) are expressed in terms of the error matrix, which is derived from the likelihood function by equations (4.33') and (4.33).

Then the error matrix E_{ij} of the parameters is

$$E_{ij} = (H^{-1})_{ij}. \tag{4.33'}$$

An example in two variables x and y will illustrate this. If the contours of ℓ are

$$\ell = -(4x^2 + y^2) \tag{4.34}$$

then ℓ is a maximum at the origin,† and

$$\ell = \ell_{max} - \tfrac{1}{2},$$

when

$$8x^2 + 2y^2 = 1. \tag{4.35}$$

† For simplicity of the algebra, we have chosen functions whose maxima are at the origin. The generalisation to functions which maximise elsewhere is straightforward.

This contour is shown in Fig. 4.6. Thus, from eqn (4.35) we see that the errors on the variables are given by

$$\left.\begin{array}{l} x = \pm\sqrt{\tfrac{1}{8}} \quad \text{(when } y = 0) \\ y = \pm\sqrt{\tfrac{1}{2}} \quad \text{(when } x = 0) \end{array}\right\} \tag{4.36}$$

or

and x and y are uncorrelated.

Our recipe (4.33') does indeed produce this result, since

$$-\frac{\partial^2 \ell}{\partial x_i \, \partial x_j} \xrightarrow{} \begin{pmatrix} 8 & 0 \\ 0 & 2 \end{pmatrix} \xrightarrow{\text{invert}} \frac{1}{16}\begin{pmatrix} 2 & 0 \\ 0 & 8 \end{pmatrix}. \tag{4.37}$$

The last matrix in (4.37) is the error matrix and hence is consistent with (4.36).

The error matrix we have obtained is identical with that of Section 3.4(a). Just as we did there, we can introduce correlations by rotating the axes (see the dashed x'- and y'-axes of Fig. 4.6). Then the likelihood function becomes

$$-\ell = \tfrac{1}{4}(13x'^2 + 6\sqrt{3}x'y' + 7y'^2) \tag{4.38}$$

and the error matrix is

$$\left(-\frac{\partial^2 \ell}{\partial x_i \, \partial x_j}\right)^{-1} = \frac{2}{64}\begin{pmatrix} 7 & -3\sqrt{3} \\ -3\sqrt{3} & 13 \end{pmatrix}. \tag{4.39}$$

This then is an example of deducing the error matrix (4.39) from the likelihood function (4.38) in which the errors on the variables x' and y' are correlated.

Of course, the likelihood function is usually determined numerically rather than analytically, and hence estimates of the second derivatives (4.33) required for the error matrix will also have to be obtained numerically.†

4.4.5 Extended maximum likelihood method

In the usual likelihood approach we determine parameters relevant to the *shapes* of distributions, but simply take the absolute normalisation as being equal to the observed number of events. However, a variant of this technique, called the 'extended maximum likelihood method' (EMLM), also provides estimates of the absolute normalisation.

The difference between the two methods is as follows. We assume that we have performed an experiment which yielded N events, characterised by observations $x_1, x_2, ..., x_N$. Usually we write down the probability, for

† We can equivalently determine the elements of the error matrix numerically from the contour $\ell = \ell_{max} - \tfrac{1}{2}$ (see p. 156).

a given set of parameters (which we wish to determine), of having obtained the observed set of x-values in a *sample of fixed size N*. In the EMLM, however, our parameter set includes one relating to the overall production rate of this type of event, and we calculate the probability (a) of observing a sample of N events, and (b) that the sample had the given x-distribution. In our maximisation procedure, we then determine all the parameters, including the one for the overall normalisation.

4.4.5.1 An angular distribution

A specific example illustrates this idea more clearly. We return to the example of an angular distribution, containing N events in all, of which F are observed in the forward hemisphere and $B(=N-F)$ in the backward one (see Section 3.2).

For the usual likelihood method, we write down the probability P_B for obtaining the observed division between forward and backward events for a sample of fixed size N, assuming that the probability of an event being produced in the forward hemisphere is f. According to the binomial distribution

$$P_B = f^F(1-f)^B N!/F!B!.$$

The maximum of $\log P_B$ with respect to f then provides us with our estimate of f as

$$\hat{f} = F/N,$$

as expected. To obtain the error σ on $\hat{f}$ we calculate

$$\sigma^{-2} = -\frac{\partial^2 \ln P_B}{\partial f^2}$$

$$= \frac{F}{f^2} + \frac{B}{(1-f)^2}$$

$$= \frac{N}{\hat{f}(1-\hat{f})}$$

where we have obtained the last result by ignoring the difference between f and $\hat{f}$. Our estimate for σ is also as expected, since the variance on the observed number of 'successes' of a binomial distribution is $Nf(1-f)$.

We can convert our estimate of f to give the estimated rates of forward and of backward events $\hat{F}$ and $\hat{B}$ as

$$\hat{F} = N\hat{f} = F$$

and

$$\hat{B} = N(1-\hat{f}) = B,$$

each with error $\sqrt{(FB/N)}$, and these errors are completely anticorrelated. We also see that the errors on F and B are smaller than $\sqrt{F}$ and $\sqrt{B}$ respectively, as expected for a binomial distribution.

In the EMLM, however, we write the probability P as the product of the Poisson probability for observing a sample of size N, and the previous binomial probability

$$P = \left\{\frac{e^{-\nu}\nu^N}{N!}\right\} \times \left\{\frac{N!}{B!\,F!}f^F(1-f)^B\right\}, \tag{3.9}$$

where ν is the extra parameter describing the expected rate at which the total number of events are produced. By maximising the logarithm of P with respect to ν and f, we obtain estimates

$$\hat{\nu} = N \pm \sqrt{N}$$

and

$$\hat{f} = F/N \pm \sqrt{[f(1-f)/N]},$$

with the errors on $\hat{\nu}$ and $\hat{f}$ being uncorrelated.

Alternatively, we can rewrite eqn (3.9) like eqn (3.10), as

$$P = (e^{-\phi}\phi^F/F!) \times (e^{-\beta}\beta^B/B!), \tag{3.10'}$$

where the parameters ϕ and β are the expected numbers of forward and of backward events. On maximising $\log P$ with respect to ϕ and β, we find, not suprisingly, that

$$\hat{\phi} = F \pm \sqrt{F}$$

and

$$\hat{\beta} = B \pm \sqrt{B},$$

again with uncorrelated errors.

The EMLM thus contrasts with the ordinary likelihood method where, because of the constraint that the total number of events was fixed at N, the *errors* on the estimates of the forward and backward numbers of events were smaller and correlated. The two approaches, however, provide identical estimates for the *number* of backward (or of forward) events.

4.4.5.2 Separating pions and kaons

A second, and more relevant, application arises when we wish to determine the particle composition of a given set of tracks. For simplicity, we will assume that our sample contains a mixture of pions and kaons only, which we are trying to distinguish. (We could, for example, be measuring the track's momentum and its energy loss dE/dx, which depends on its speed and hence on its mass.)

In the ordinary likelihood approach, we would maximise log $\mathscr{L}$, where

$$\mathscr{L} = \prod \{f_\pi \exp(-(x_i - x_\pi)^2/2\sigma^2) \\ + (1 - f_\pi)\ \exp(-(x_i - x_K)^2/2\sigma^2)\}.$$

Here x_π and x_K are the expected positions of the pion and kaon peaks for the variable we are using to distinguish the particles (e.g. dE/dx); $x_i \pm \sigma$ is the measured value for the ith track; and f_π is the fraction of pions that we wish to determine. The product extends over the N tracks of the sample, and we have omitted irrelevant constants from our definition of $\mathscr{L}$.

The *extended* likelihood function $\mathscr{L}_E$ includes an extra factor for the probability of obtaining a sample of size N from a Poisson distribution of mean ϕ i.e.

$$\mathscr{L}_E = \frac{e^{-\phi}\phi^N}{N!}\mathscr{L}.$$

Our variables are now f_π and ϕ, or equivalently $\phi_\pi = \phi f_\pi$ and $\phi_K = \phi(1 - f_\pi)$, the numbers of pions and kaons respectively. (In the ordinary likelihood approach, our estimate of the number of pions is Nf_π.) In terms of ϕ_π and ϕ_K, the extended likelihood can equivalently be written as

$$\mathscr{L}_E = \frac{e^{-(\phi_\pi + \phi_K)}}{N!}\prod\{\phi_\pi \exp(-(x_i - x_\pi)^2/2\sigma^2) \\ + \phi_K \exp(-(x_i - x_K)^2/2\sigma^2)\}.$$

In estimating the numbers of pions and kaons, three different factors are relevant:

(i) How much ambiguity is there in resolving pions and kaons experimentally? This depends crucially on the magnitude of $(x_\pi - x_K)/\sigma$.

(ii) What is our estimate of the number of pions in *our particular sample*? This is answered by the maximum likelihood approach.

(iii) What is our estimate of the number of pions for *experiments like ours*? Here the extended likelihood method is needed.

Both (ii) and (iii) would include in their error estimates the effects of any identification problems as mentioned in (i). The magnitude of this is most readily obtained as $\sqrt{(-C)}$, where C is the covariance of the estimates of ϕ_π and ϕ_K in the EMLM; in the absence of ambiguities, C is zero.

Whether we should use the ML or the EML approach depends on the question we are trying to answer. If, for example, the tracks are the decay products of some resonance, and we want to determine its branching ratios, then the ML method is required. If, however, we are determining its partial decay rates, then we need the EMLM.

Thus if we observe 100 tracks with 96 and 4 of them in well-separated pion and kaon peaks respectively, the fraction of pions is determined as 0.96 ± 0.02. The estimate of the number of pions expected in repeated experiments of this sort, however, is 96 ± 10 since the constraint of 100 tracks overall would not apply (i.e. we are taking into account the fact that the total number of tracks would have statistical variations from experiment to experiment).

Finally, we note that this problem of distinguishing pions from kaons would be formally identical to our earlier example of classifying tracks as forwards or backwards provided that we modified it by assuming that there was a finite resolution to the measured track angle, which could produce an ambiguity in their assignment to the two hemispheres.

4.5 Least squares

4.5.1 What is it?

Let us assume that we are given an experimental distribution. In order to fix our ideas, we shall take this to be a histogram (i.e. the number of observed events plotted as a function of the relevant variable), although almost everything in this section is equally applicable to any set of observations $y_i \pm \delta y_i$ at known, varying x_i (e.g. the length of a bar of metal at several known temperatures).

We denote the entries by y_i^{obs}, where i labels the particular bin of the histogram. As usual, we are trying to fit this data by a functional form $y_i^{th}(\alpha_j)$, where the α_j are parameters in the theoretical expression. We then construct

$$S = \sum_{i-1}^{bins} \left(\frac{y_i^{obs} - y_i^{th}(\alpha_j)}{\sigma_i} \right)^2 \tag{4.40}$$

where the σ_i are some errors, to be discussed below. If the theory (with some particular values of the parameters) is in 'good' agreement with the data, then y^{obs} and y^{th} do not differ by 'much' and hence the value of S will be 'small'. Fortunately it is easier to quantify how small 'small' should be in this case than in the corresponding problem of the maximum likelihood method; we return to this point in Section 4.6.

What is σ_i? It is supposed to be the error on the *theoretical* prediction (i.e. if we predict that a particular bin should contain 15.3 events, then if the experiment were repeated many times we would expect the observed numbers of events in that bin to follow a Poisson distribution with an error σ of $\sqrt{15.3}$).

Two extreme examples illustrate why we should use the error on the theoretical estimate (rather than that estimated from the observed number).

Example (i)
If $y^{obs} = 1 \pm {}^{\prime}1{}^{\prime}$ and $y^{th} = 0(\pm 0)$, then the theory is clearly wrong since we have indeed seen something when we expected nothing. Then S will be $((1-0)/0)^2$ which is suitably large, and indicates that the theory is to be rejected.

Example (ii)
If $y^{obs} = 0 \pm {}^{\prime}0{}^{\prime}$ and $y^{th} = 1 \pm 1$, we should not be too surprised since for a Poisson distribution with a mean of unity, there is a 37% probability of observing 0 events. In this case S will be $((0-1)/1)^2$ which is suitably small, to indicate that there is no inconsistency between theory and experiment.

Had we taken σ_i as the error estimated from the observed number of events, then the values of S in the two examples would have been interchanged, and our conclusions would have been incorrect. (Compare the discussion in Comment (ii) of Section 1.6 on the dangers of using estimated errors rather than their 'correct' values.)

Having justified why we should use the *theoretical* error for σ_i, we now explain why it is common practice to use the estimates based on the *observed* number of events y_i^{obs}.

Excuse (i)
The minimisation of the function S of eqn (4.40) with respect to the parameters α_j is much simpler if the σ_i are independent of α_j. This would not be the case if we used the error on the theoretical estimate which is the square root of yth and hence involves α_j. But with σ_i taken simply as the error estimated from the observed number of events, then the α_j occur only in y^{th} in the numerator.

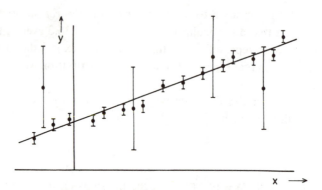

Fig. 4.7. A least squares straight line fit to some data points. Four of the points have considerably larger errors than the others, but by weighting each point in the fit inversely as the square of the experimental error σ_i, the less well measured points do not significantly pollute the better data, while still not being completely ignored in the fit. In real life, the errors on the experimental points are likely to have a continuous range in magnitude.

Excuse (ii)

It is very much easier to combine data of different basic accuracies. Thus in Fig. 4.7, some of the points have much larger error bars than the others. When, say, a straight line is fitted to these data points, the use of the experimental errors for σ_i enables the line to give less significance to the points of lower accuracy.

Excuse (iii)

We do not want σ to be too small since we later wish to use the approximation that the Poisson distribution resembles the Gaussian one (see Section 4.6). To avoid this happening, it may be necessary to combine adjacent bins in order to make the number of events and σ large enough. If σ_i is a function of the α_j, then during the minimisation process, a particular bin may keep moving backwards and forwards across the boundary defining the acceptable minimum value of σ. Then the total number of bins used keeps changing; this is clearly undesirable.

A byproduct of combining bins is that the perils involved in using the experimental errors described in the examples above are made far less serious.

To demonstrate how the method is used for determining parameters, we return to our previous example of an angular distribution which we are attempting to describe by the expression (4.1). From the experimental data, we first construct a histogram. We have, of course, to choose a suitable bin size, which should be neither too small (as this would result in too few events in each bin) nor too large (as we then lose resolution in determining the functional form). We then calculate S as a function of our parameter b/a, and choose the smallest value† of S; the corresponding value of b/a is our best estimate of this parameter. The error on b/a is given either by calculating $(\frac{1}{2}d^2S/dp^2)^{-\frac{1}{2}}$, or by seeing how much b/a must be altered in order to make S increase from S_{min} to a value of $S_{min}+1$. For a problem with several parameters, their best values are given by $\partial S/\partial p_i = 0$, and their error matrix by the inverse of $[\frac{1}{2}\partial^2 S/\partial p_i \partial p_j]$.

To appreciate that this procedure gives sensible results, we note that:

Comment (i)
If we have only a single measurement $(y^{obs} \pm \sigma)$ of a particular quantity, then to obtain our best estimate y^{th} of this quantity, we construct

$$S = \left(\frac{y^{obs}-y^{th}}{\sigma}\right)^2.$$

This minimises with $S = 0$ for $y^{th} = y^{obs}$; and $S = 1$ when $y^{th} = y^{obs} \pm \sigma$. This is eminently reasonable.

Comment (ii)
If we have two measurements y_1 and y_2 of a single quantity, and their errors are equal, then to deduce the best estimate y^{th}, we construct

$$S = \left(\frac{y_1-y^{th}}{\sigma}\right)^2+\left(\frac{y_2-y^{th}}{\sigma}\right)^2.$$

A little trivial algebra enables us to deduce that our best estimate for y^{th} is $\frac{1}{2}(y_1+y_2)$ (as obtained from $dS/dy^{th} = 0$) and its error is $\sigma/\sqrt{2}$ (deduced from $S = S_{min}+1$). Again this is exactly what we would have expected (compare Section 1.6).

Comment (iii)
In the case when the least squares method is equivalent to the maximum likelihood method (see Comment (vi) of Section 4.4.3), they will give the same answer to a particular problem.

---

† It is easiest to visualise the process by assuming that we have in fact calculated S for a large series of b/a values, and then simply choose the smallest value by inspection. In our example, however, the fitted function $1+b/a \cos^2 \theta$ is *linear* in the *parameter* b/a; for such cases, there is a simple analytic method for finding the minimum. This is illustrated in Section 5.1.

4.5.2 Notes on the least squares method

Note (i)

As already mentioned, before we can start we must choose a suitable bin size in which to plot our histogram. Hopefully the results will depend but little on the particular choice of bin size.

It is not imperative that all the bins should be the same size. The most compelling reason for avoiding varying bin size is that we then are faced with a further question on how to choose all the bins; such arbitrary choices are best avoided.

Note (ii)

It is undesirable to have less than five events in any bin. This is because it is useful to be able to assume that the errors are Gaussian distributed (see Section 4.6), and the actual Poisson distribution approximates to a Gaussian only asymptotically (which is conventionally taken to be for a mean of five or larger). A second reason is that as we are using the experimental error for σ_i (rather than the error on the theoretical prediction), we want to avoid the unfortunate situations that can occur when the numbers of events are very small (see Examples (i) and (ii) in Section 4.5.1 above).

To avoid having bins with few events, such bins should not be ignored, but should be grouped with adjacent bins in order to bring the numbers up to at least five. In principle any bin in the histogram can be added to any other, but it is generally sensible to combine only adjacent bins.

It is, of course, essential that when experimental bins are combined, the theoretical estimates for the corresponding bins are combined too.

Note (iii)

So far, we have discussed fitting a distribution in only one variable (although the theoretical distribution can contain several parameters). The least squares method can fairly easily be extended to dealing with distributions in two variables. These could be, for example, a Dalitz plot, or a distribution of arrows shot at a target as a function of vertical and horizontal displacements from the bull's eye (see Fig. 4.8). This problem is reduced to the 1-dimensional case by superimposing a grid over the distribution, and then reading off the number of events in each bin in some sequence. For each set of parameters, the theoretical distribution is then compared with these experimental numbers in the same sequence. As in the 1-dimensional case, it may be necessary to combine a sparsely populated

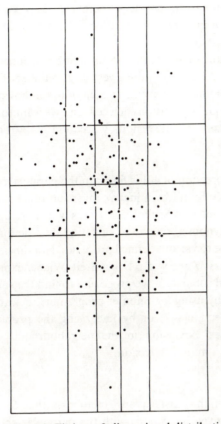

Fig. 4.8. Fitting a 2-dimensional distribution by the least squares method. The points represent the arrival positions of arrows shot at a target; the vertical spread is larger than the horizontal one because of the effect of gravity. The method consists in first superimposing a grid over the experimental points, and reading off the number of points per cell in some specified order. The grid size can usefully be constructed smaller in the central region of higher point density, and larger at the edges. There is no particular reason for choosing the x and the y lines identically. (Indeed they cannot be if the two variables are physically different.) A cell with few points inside should be combined with an adjacent cell. Then the chosen functional form is used to fit the data. For our example, this could consist of a 2-dimensional Gaussian distribution, with the means and variances (and perhaps the correlation term) as variables. The theoretical prediction for each cell is then compared with the experimentally observed number by use of the usual expression for S, where the summation extends over all the cells covering the 2-dimensional distribution.

bin with one of its neighbours. If the density of points varies considerably over the region of interest, it may well be desirable to have a variable grid size.

In principle, this procedure can be extended to any number of dimensions, but as this number increases, we will require a very large number of events in order to keep a sensible number of events in each of our n-dimensional cells. Thus if we require about 20 events in each cell, and we want to divide each of our n variables into ten separate regions, we require 20×10^n events.

Note (iv)

When the function y in expression (4.40) is linear in the parameters, then the minimum of S is readily found by a matrix inversion procedure. An example of this is given in detail in Section 5.1. Cases which are linear in the parameters include polynomial functions, or fitting the fractions of a given reaction which are due to various defined processes. Non-linear cases (for example, fitting the mass M and width Γ in a Breit–Wigner expression for a resonance shape) need some iterative procedure to find the solution. This could be done either by using a computer programme to search for the minimum of S (see Section 4.7) or by linearising the problem by performing a Taylor series expansion in the neighbourhood of the minimum, and then doing a matrix inversion.

Note (v)

The most useful feature of the least squares method is that S_{min} is a measure of how well the given hypothesis with the particular choice of parameters fits our data; a small value of S_{min} implies better agreement (see Section 4.6).

However, in many real life situations (for example, those concerning physics), the experimental data consist of many (perhaps even an infinite number of) distributions, all of which may in principle be predicted by the theory.† The problem is how many and which of these should be used in the comparison of theory and data, since not all the variables are independent.

The situation is aggravated by the fact that even with, say, three

---

† In principle the problem can be solved as follows. Each individual event can be characterised by a certain number n of independent variables. We can then plot each event in the relevant n-dimensional space, and by constructing an n-dimensional grid, can then use the method described in Note (ii) to compare any theory with experiment. The limited amount of experimental data usually makes this method impractical, so instead one resorts to comparing theory and experiment for a series of 1-dimensional distributions which one hopes are more or less independent.

variables that are related to each other (for example, direction cosines which are constrained by

$$\cos^2 \alpha + \cos^2 \beta + \cos^2 \gamma = 1),$$

knowledge of the $\cos \alpha$ and $\cos \beta$ distributions does not allow us to calculate that for $\cos \gamma$. Thus even though a theory may be consistent with $\cos \alpha$ and $\cos \beta$ distributions, it could still get the $\cos \gamma$ one badly wrong. So should we include all three distributions in our definition of S?

The choice of a sensible set of distributions to use in such a method is a matter of experience and judgement. Those distributions which are most sensitive to the various theories are clearly going to be most useful in distinguishing between them. But the use of several distributions in variables which are not independent will in general mean that S does not display its usual quantitative properties; thus the error estimates are liable to be incorrect and statements about confidence levels concerning agreement between theory and data (see Section 4.6 below) may also be wrong.

4.5.3 *Correlated errors for* y_i

Finally we consider the modification necessary in the definition of S when the errors of the various y_i are correlated. As in Section 3.4, we deduce the formula to be used in the correlated case by starting from the uncorrelated situation with which we are familiar, and introduce the correlations by rotating the co-ordinates. Thus for two uncorrelated variables z' and y', we define

$$S = \left(\frac{z' - z_0'}{\sigma_{z'}}\right)^2 + \left(\frac{y' - y_0'}{\sigma_{y'}}\right)^2, \tag{4.41}$$

where $z' \pm \sigma_{z'}$ and $y' \pm \sigma_{y'}$ are the measured quantities, and z_0' and y_0' the predicted ones.

We rotate the measured and the predicted co-ordinates by the transformation

$$\left.\begin{aligned} z' &= z \cos \theta - y \sin \theta, \\ y' &= z \sin \theta + y \cos \theta. \end{aligned}\right\} \tag{4.42}$$

This implies that the errors transform as

and
$$\left.\begin{aligned} \sigma_{z'}^2 &= \sigma_z^2 \cos^2 \theta + \sigma_y^2 \sin^2 \theta - 2 \operatorname{cov}(z, y) \sin \theta \cos \theta \\ \sigma_{y'}^2 &= \sigma_z^2 \sin^2 \theta + \sigma_y^2 \cos^2 \theta + 2 \operatorname{cov}(z, y) \sin \theta \cos \theta. \end{aligned}\right\} \tag{4.43}$$

Table 4.2. *Comparison of various techniques for parameter determination*

	Moments	Maximum likelihood	Least squares
How easy?	Very, provided suitable moments can be found	Normalisation and maximisation can be messy	Needs minimisation
Efficiency	Not very	Usually most efficient	Sometimes equivalent to max. like.
Input data	Individual events	Individual events	Histograms
Estimate of goodness of fit	Rather messy to obtain	Very difficult	Easy
Constraints among parameters	Cannot be imposed	Easy	Can be imposed
Are n-dimensional problems difficult?	Not if suitable moments can be found	Normalisation and maximisation get messier	Problem of which distributions to choose
Weighted events	Easy	Can be used	Easy
Overflow	Excluded	Included	Included
Background subtraction	Easy	Can be troublesome	Easy
Error estimate	From spread of individual values	$\left(-\dfrac{\partial^2 \ell}{\partial p_i\,\partial p_j}\right)^{-\frac{1}{2}}$	$\left(\dfrac{1}{2}\dfrac{\partial^2 S}{\partial p_i\,\partial p_j}\right)^{-\frac{1}{2}}$

Furthermore since by assumption the errors on z' and y' are independent,

$$\operatorname{cov}(z', y') = (\sigma_z{}^2 - \sigma_y{}^2)\sin\theta\cos\theta + \operatorname{cov}(z, y)(\cos^2\theta - \sin^2\theta)$$

$$= 0. \tag{4.44}$$

On inserting eqns (4.42) and (4.43) into (4.41), and making use of the condition (4.44), after a lot of algebra we eventually obtain

$$S = \frac{1}{\sigma_z{}^2\sigma_y{}^2 - \operatorname{cov}(z,y)^2}\big[\sigma_y{}^2(z-z_0)^2 + \sigma_z{}^2(y-y_0)^2$$

$$-2\operatorname{cov}(z,y)(z-z_0)(y-y_0)\big]$$

$$= H_{11}(z-z_0)^2 + H_{22}(y-y_0)^2 + 2H_{12}(z-z_0)(y-y_0),$$

where H_{11}, H_{22} and H_{12} are the elements of the inverse error matrix, i.e.

$$\begin{pmatrix} H_{11} & H_{12} \\ H_{21} & H_{22} \end{pmatrix}^{-1} = \begin{pmatrix} \sigma_z^{\,2} & \mathrm{cov}\,(z, y) \\ \mathrm{cov}\,(z, y) & \sigma_y^{\,2} \end{pmatrix}.$$

For several correlated variables, we can write S in matrix notation as

$$S = \Sigma_{ij}\, \tilde{\Delta}_i\, H_{ij}\, \Delta_j,$$

where the vector

$$\Delta_j = y_j^{\,obs} - y_j^{\,th}$$

and H_{ij} is the inverse error matrix of the observed values y^{obs}. The formula (4.40) given earlier is just a special case of this, when the error matrix is diagonal.

4.6 Hypothesis testing

4.6.1 *Use of weighted sum of squared deviations*

We now return from the simpler question of what are the best values of the parameters, to the more fundamental one of whether our hypothesis concerning the form of the data is correct or not. In fact we will not be able to give a 'yes or no' answer to this question, but simply to state how confident we are about accepting or rejecting the hypothesis.

In simple cases, the hypothesis may consist simply of a particular value for some parameter. For example,

(i) Is the velocity of light the same today as it was last year?

(ii) Does the value of the polarisation of the electrons emitted in a particular β decay process agree with the prediction of the $V - A$ theory?

(iii) Is a particular angular distribution isotropic? (i.e. Is b/a of eqn (4.1) equal to zero?)

In these cases, we simply use any of the parameter determining methods already described to obtain the value of the parameter and its error, and then compare it with our hypothesised value. Then using Fig. 3.4, we will be able to make a numerical statement on the probability of obtaining a result equal to ours (or worse), assuming that the hypothesis is true. If this probability is very low, then we will reject this particular hypothesis.

In general it is preferable, however, to perform *distribution* testing rather than parameter testing. Thus in the Example (iii) above, it is more sensible

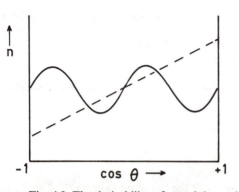

Fig. 4.9. The desirability of examining a distribution rather than simply determining a parameter when we are hypothesis testing. If we fit either the solid or the dashed distribution in $\cos \theta$ by an expression $[1 + b/a \cos^2 \theta]$, the value of b/a is liable to be close to zero. This does not imply that either distribution is isotropic.

to check whether the distribution of $\cos \theta$ is consistent with being flat, rather than simply testing whether b/a is consistent with zero, since there are many non-isotropic distributions which would give such a value of b/a if we simply tried to fit the data by expression (4.1); a couple of examples are shown in Fig. 4.9.

Distributions are tested by the χ^2-method. When the experimentally observed numbers of events y_i^{obs} in each bin† are Gaussian distributed with mean y_i^{th} and with variance σ_i^2, the S defined in eqn (4.40) is distributed as χ^2. So in order to test a hypothesis we

  (a) construct S and minimise it with respect to the free parameters;
  (b) determine the number of degrees of freedom v from

$$v = b - p,$$

  where b is the number of bins of the distribution included in the summation for S, and p is the number of free parameters which are allowed to vary in the search for S_{min}; and
  (c) look up in the relevant set of tables the probability that, for v degrees of freedom, χ^2 is greater than or equal to our observed value S_{min}.

Some χ^2 distributions, which depend on the number of degrees of

---

† As at the beginning of Section 4.5.1, most of the remarks here apply to any type of distribution, and not just to histograms.

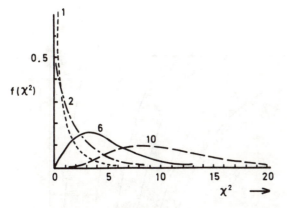

Fig. 4.10. χ^2-distributions, for various numbers of degrees of freedom v (shown by each curve). As v increases, so do the mean and variance of the distribution.

freedom, are shown in Fig. 4.10. They have the property that the expectation value

$$\overline{\chi^2} = v \tag{4.45}$$

and the variance

$$\sigma^2(\chi^2) = 2v. \tag{4.45'}$$

Thus large values of S_{min} are unlikely, and so our hypothesis is probably wrong. (Similarly, very small values of S_{min} are also unlikely, and so again something is suspicious – cf. Section 3.3.)

More useful than the χ^2 distribution itself is

$$F_v(c) = P_v(\chi^2 > c), \tag{4.46}$$

i.e. the probability that, for the given number of degrees of freedom, the value of χ^2 will exceed a particular specified value c. Such distributions are available in almost all books on statistics; some are shown in Fig. 4.11. The relationship between the χ^2 distribution and that of F is analogous to that between the Gaussian distribution and the fractional area in its tails.

What does F mean? If our experiment is repeated many times, and assuming that our hypothesis is correct, then because of fluctuations we will get a larger value of S_{min} than the particular one we are considering (i.e. a worse fit) in a fraction F of experiments. (The interpretation is thus analogous to that for the case of comparing a measured quantity whose error is known with some standard value – see Section 3.3.)

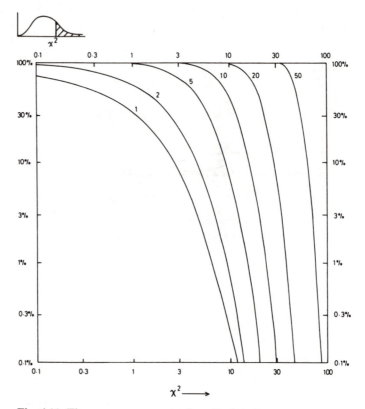

Fig. 4.11. The percentage area in the tail of χ^2-distributions, for various numbers of degrees of freedom, shown by each curve. Both scales are logarithmic. These curves bear the same relation to those of Fig. 4.10, as does Fig. 3.4 to the normal distribution of Fig. 1.5.

For example, in a $\cos \theta$ histogram, let us assume that there are 12 bins and that when we fit the expression $N(1 + b/a \cos^2 \theta)$ to the data, we obtain a value of 20.0 for S_{min}. In this case we have ten degrees of freedom (12 bins less the two parameters N and b/a). From Fig. 4.11, we see that the probability of getting a value of 20.0 or larger is about 3%.

As usual, it is up to us to decide whether or not to reject the hypothesis as false on the basis of this probability estimate but at least we have a numerical value on which to base our decision.

In deciding whether or not to reject a hypothesis, we can make two sorts of incorrect decision.

(a) Error of the first kind

In this case we reject the hypothesis H when it is in fact correct. This should happen in a well known fraction F of the tests, where F is determined (from Fig. 4.11) by the maximum accepted value of S_{min}. But if we have biasses in our experiment so that the actual value of the answer is incorrect, or if our errors are incorrectly estimated, then such errors of the first kind can happen more (or less) frequently than we expect on the basis of F.

How serious an error of the first kind is depends on whether our hypothesis applies to the experiment as a whole, or whether it is used to test lots of different sets of data in a consistency check (see the examples in Section 4.6.2 below).

The number of errors of the first kind can be reduced simply by increasing the limit on S_{min} above which we reject the hypothesis. The only trouble is that this is liable to increase the number of errors of the second kind, and so some compromise value of the limit must be chosen.

(b) Error of the second kind

In this case we fail to reject the hypothesis H when in fact it is false, and some other hypothesis is correct. That is, the value of S_{min} accidentally turns out to be small, even though the hypothesis H (i.e. the theoretical curve y^{th} that is being compared with the data) is incorrect. In general, it is very difficult to estimate how frequent this effect is likely to be; it depends not only on the magnitude of the cut used for S_{min} but also on the nature of the competing hypotheses. If these are known, then we *may* be able to predict what distributions they will give for S_{min} (where S_{min} is calculated on the assumption that the original hypothesis is correct), and hence how often we will be incorrect in accepting H.

4.6.2 Types of hypothesis testing

The hypothesis we are testing may relate to the experiment as a whole, or alternatively it may be used simply as a selector for subsets of data samples which satisfy specific criteria. We now discuss these two possibilities in turn.

(a) Hypothesis relates to whole experiment

We observe an angular distribution from the decay of a resonance, and want to know: 'Does the resonance have spin zero?' This would imply that the angular distribution is isotropic.

In testing such a hypothesis, which relates to the experimental data as

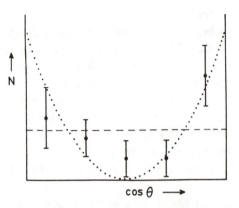

Fig. 4.12. The angular distribution for the decay of a state whose spin we wish to determine. If the spin is zero, the decay distribution must be isotropic (dashed line). We calculate the value of S_{min} for this hypothesis. As we have five experimental points, there are four degrees of freedom for this hypothesis, since the only variable is the normalisation. Then if S_{min} is larger than 10, we would reject this hypothesis, since the probability that χ^2 for four degrees of freedom exceeds 10 is only 5%. (Choosing this value of S_{min} implies that in 5% of our experiments we would reject a spin assignment when it is in fact correct – an error of the first kind.) In our case, S_{min} is 8.7, so the hypothesis is not rejected.

   This, however, does not necessarily mean that the spin is zero. If it were 1, the predicted decay distribution may be $\cos^2 \theta$ (dotted curve). The value of S_{min}' for this hypothesis is 4.1, which is also below our rejection cut. The errors on our data are so large that we have poor discrimination between these two hypotheses. Had we simply accepted the hypothesis of spin zero when spin one was the correct value, we would have made an error of the second kind.

a whole, an error of the first kind is serious and in this example so is an error of the second kind; in the former case, we reject the spin zero case when it is in fact true, while in the latter we accept it when the spin is non-zero, i.e. we simply get the answer wrong.

   As usual we should be able to estimate how often we will make an error of the first kind from the magnitude of the cut on S_{min} that we use for testing the hypothesis of a flat angular distribution. In this experiment, the alternative hypotheses are well defined: if the spin is not zero, it is 1, 2, 3, ... (in units of $\hbar$). It may also be possible to calculate what angular distributions we expect for each of these cases, and hence we can deduce how often each of these would give a low value for S_{min} (where S_{min} is calculated assuming the distribution is flat). This then gives us an estimate of the frequency for making errors of the second kind.

Apart from the magnitude of the S_{min} cut, it will depend on how different are the predicted angular distributions and on the accuracy of the data; the bigger the errors on the experimental points (because not many events have been observed in the decay distribution and/or because there are large corrections for detection inefficiencies), the poorer will be the experiment in discriminating among the different spin possibilities, and the *power* of the test is said to be low (see Fig. 4.12). Where relevant, it is important to state in the conclusions of such an experiment that although the data may be consistent with the given hypothesis, the experiment is not good enough to rule out various other possibilities.

In other cases we may simply be testing whether some theory or hypothesis is correct, e.g. is our experiment consistent with Einstein's general theory of relativity, or at a more mundane level, are our results consistent with those obtained by Jones and Smith last year? In these cases an error of the second kind merely means that we fail to reject the hypothesis when it is in fact incorrect. In the 'Einstein' example, this does not result in us making a mistake, but we do fail to achieve fame by discovering that Einstein may have been incorrect.

With respect to errors of the first kind, our choice of what cut-off value for F would be reasonable will be influenced by the type of hypothesis we are testing. Thus if it is simply checking on Jones and Smith, we would probably be satisfied with a 95% confidence cut, i.e. we would not be too worried by incorrectly claiming that they were wrong in one out of 20 times we checked such experiments. On the other hand, with a result as dramatic as showing that Einstein's theory is incorrect, we would doubtless feel that we needed a much more convincing demonstration (not more than, say, 1 in 1000 chance† of a statistical fluctuation) before we would be prepared to risk our reputation with such a claim.

(b) Hypothesis used as data selector
An experiment may consist of a large set (say 50000 examples) of interactions of a beam of protons with a hydrogen target, in each of which four charged tracks are observed and measured. (See Fig. 4.13.) We are

---

† This corresponds to a 3 to 4 standard deviation effect. As mentioned earlier, a serious problem here is that we may have underestimated our errors, and in fact the discrepancy may be only at the ~ 2 standard deviation level, whose significance is much lower (~ 1 in 20).

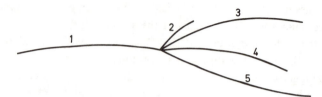

Fig. 4.13. An example of an interaction producing four charged particles, and an unknown number (possibly zero) of neutrals. The charged particles leave visible tracks in detectors such as the bubble chamber; the neutrals are assumed to be undetected. A magnetic field makes the charged tracks move in helices; the radius of curvature is proportional to the particle's momentum divided by its charge. Thus the incident particle 1 and the charged outgoing particles 2–5 have their directions and momenta measured; the possible neutrals are unmeasured. This event could be an example of the reaction pp → ppπ⁺π⁻. By using the observed track variables, we can test this hypothesis. By obtaining a large sample of examples of this reaction, we can investigate physics topics of interest e.g. whether resonance production is important, what reaction mechanisms are relevant, etc.

interested in testing the hypothesis that these interactions are examples of the reaction

$$\text{pp} \rightarrow \text{pp}\pi^+\pi^-. \tag{4.47}$$

The hypothesis is tested by seeing whether the measured directions and momenta of the tracks are consistent with those expected for reaction (4.47) on the basis of energy and momentum conservation; a simplified case of such a kinematic fitting problem is described in Section 5.2.2.

Here we are using our hypothesis to check the individual sets of data, and to provide us with a hopefully pure sample of examples of reaction (4.47), which we then wish to study with a view to extracting some interesting physics. Errors of the first kind simply correspond to rejecting a small fraction of genuine examples of reaction (4.47) which happen to give a value of S_{min} above the cut we are using. This should not be too serious; the reduction is the size of the data sample due to the rejection of these events should be small, and its magnitude should be calculable (simply from the value of the S_{min} cut), so that we can correct for this loss. The only danger is that we reject events preferentially from some specific kinematic configuration (for example, where one of the tracks has a very high momentum), and hence the events which are accepted as examples of reaction (4.47) constitute a biassed sample. All these problems are reduced by choosing a cut on S_{min} which is not too low; it should remove not more than about 20% of the genuine events.

As a check that our selection criteria are working correctly, we can plot the distribution of probabilities ($P_\nu(\chi^2 > c)$ of eqn (4.46)) for the accepted events. This should be flat from some minimum probability value (corresponding to the maximum accepted χ^2) up to unity. The degree of uniformity of this distribution is a sensitive test that, for example, the momenta and angles are not biassed, and that their errors are correctly estimated. In general contaminations from other reactions (see below) will produce a peak or enhancement at the low probability end of the plot.

Errors of the second kind correspond to accepting events as examples of reaction (4.47) when they in fact are produced by some other reaction with four visible charged tracks, e.g.

$$pp \to pp\pi^+\pi^-\pi^0, \tag{4.48a}$$

$$pp \to pn\pi^+\pi^+\pi^-, \tag{4.48b}$$

$$pp \to ppK^+K^-, \tag{4.48c}$$

$$pp \to pp\mu^+\mu^-. \tag{4.48d}$$

If the beam contains a significant admixture of π^+, then other possibilities are

$$\pi^+p \to p\pi^+\pi^+\pi^-, \tag{4.49a}$$

$$\pi^+p \to p\pi^+\pi^+\pi^-\pi^0, \tag{4.49b}$$

$$\pi^+p \to n\pi^+\pi^+\pi^+\pi^-, \tag{4.49c}$$

etc.

Thus errors of the second kind constitute a potentially more dangerous problem; our data sample is contaminated. The extent of this contamination is difficult to estimate. It will depend on how frequently the reactions (4.48) and (4.49) produce kinematic configurations resembling those of reaction (4.47) and also on the relative production cross-sections of the various contamination reactions compared with our reaction (4.47). Other factors affecting the magnitude of these errors are

(a) how well the events are measured; if the tracks have poorly determined momenta, then it will be difficult to distinguish the different reactions; and

(b) the effective number of constraints, i.e. the number of independent checks it is possible to perform corresponding to each hypothesis (see Appendix 2).

For example, since the μ mass is very close to that of the π, reaction (4.48d) will be very difficult to distinguish from reaction (4.47) simply on the basis of measurements of directions and momenta. But since the rate of reaction (4.48d) is very small (down by a factor of about 10^{-4}) compared with that of reaction (4.47), the overall number of events produced by reaction (4.48d) which are accepted as examples of reaction (4.47) will be small. Similarly, reaction (4.49a) may not be readily distinguishable from (4.47) without particle identification, but again the effect will not be too serious for a relatively pure beam of incident protons. On the other hand, reaction (4.48a) will be consistent with (4.47) in only a small fraction (say 1–5%) of cases, but since its production rate may be comparable to or larger than that for (4.47), the contamination in this case is more serious.

These contaminations are in general reduced by lowering the value of the cut on S_{min}. The problem thus reduces to choosing a value for this cut which will provide a suitable compromise between the problems associated with errors of the first and second kinds.

This type of example is considered again in greater detail in Section 5.2.

4.7 Minimisation

Both the least squares and the maximum likelihood methods require that we minimise a function (S and $-\ell$ respectively) of several variables. There are several approaches to this:

Approach (i)
When S is linear in the parameters, we can find the minimum explicitly by a matrix inversion process. (An example of this is given in Section 5.1.) Some likelihood methods also have analytic solutions.

Approach (ii)
For a non-linear problem with only one variable, it is best to write our own computer program for calculating S at several different values of the single variable. Such a problem could involve, for example, trying to find the best value of the central mass M_0 for a fit of a Breit–Wigner formula (eqn (4.20)) with known Γ to data on a resonant state. Then we look at the values of S as a function of M_0 in order to select the best value and its error (from the condition that S increases from S_{min} to $S_{min}+1$). It is more useful, however, to have a computer routine which inspects the set of values of S to find the three smallest values, and calculates the parabola through them in order to find the position of the minimum and its error automatically (see Fig. 4.14).

This approach can be extended to problems with two variables (for

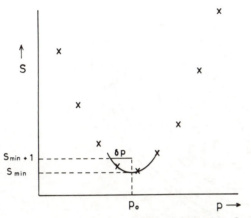

Fig. 4.14. Do-it-yourself minimisation, for a problem involving a single parameter p. The sum of squares S is calculated at a series of values of the parameter p; these are represented by the crosses. The minimum is then obtained either by inspection, or by a simple computer subroutine which calculates the parabola (solid curve) through the three points nearest the minimum. The minimum of the parabola gives the best estimate p_0 of the parameter, and the width of the parabola determines the error δp (assumed to be symmetric).

example, fitting both the mass M_0 and width Γ of the Breit–Wigner expression), by calculating S over a 2-dimensional grid of values for M_0 and Γ. Then our parabolic routine must be replaced by one which uses six points near the minimum in order to determine the best values and the errors of both the parameters, and the correlation between them (see Fig. 5.8).

Approach (iii)

For more complicated situations, the above approach can in principle be extended but comes increasingly tedious. The problem is that in the presence of correlations, it is insufficient to try to minimise a function with respect to each of its n independent variables one at a time in turn; such a procedure may not converge in any reasonable number of iterations on the true minimum (see Fig. 4.15).

It is thus advisable to use a general minimisation program, of which many variants exist. This is because we are not unique in wanting to minimise a function, and whatever the specific details of our problem, it is highly likely that this type of situation has occurred previously. Hence many such programs, which use a variety of techniques to search for a minimum, already exist.

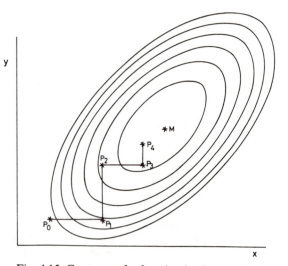

Fig. 4.15. Contours of a function in the x–y plane. We are trying to find the minimum M, by starting from an arbitrary position P_0 and then varying x and y in turn, at each stage minimising the function with respect to that single variable. The search progresses through the points $P_1 P_2 P_3 \ldots$. Because of the strong correlation between x and y (i.e. the value of ρ is large), the convergence towards the minimum becomes slow.

Basically these programs evaluate the function at a variety of locations, and then adopt some specific strategy in order to deduce a next approximation to the minimum. A feature of these programs is that it is not necessary† for the user to be aware of these details of the minimisation techniques that are employed.

Our first task is to select the most suitable minimisation program from among those that exist on the library of our local computer. Our choice will depend on whether we have a preference for a particular minimisation technique, whether we are going to provide the program with analytic expressions for the derivatives of the function or whether the program will have to calculate them numerically, whether the variables can assume any

---

† Although it is not necessary to know these details when we are finding the minimum in a relatively simple problem, for more complicated cases it is desirable to have some familiarity with the way the program works. This will enable us to have confidence (or lack of it) that the answer printed out at the end of the calculation is *the* minimum of the function, rather than some local depression into which the program has dropped and from which it cannot get out; that the quoted errors on the best values of the parameters are reasonable; etc.

values or whether they are to be confined to limited ranges (e.g. -1 to $+1$, or zero to infinity), whether there are constraints among the variables, etc.

All that is necessary in using such a program is to write a subroutine which defines the function which is to be minimised (and perhaps also its derivatives); and to provide some initialisation information that the program requires (e.g. how many variables the function uses, the range of each variable to be examined, how much print-out we want, perhaps some parameters which define when the program has got close enough to the minimum for it to stop, whether we require the error matrix for the parameters, etc.). These programs usually also provide the user with the option of including starting and finishing routines, which can respectively be used, for example, for reading in data to define the function to be minimised, and to print out information relevant to the length of the run.

Since some degree of familiarity with such a program is desirable, if this is the first time that this particular minimiser is to be used, it is prudent to try it out on some simple situation, before our real problem is attacked in earnest. Thus we could set the program to minimise $(x-a)^4+(y-b)^2$.

Approach (iv)

For minimisation problems with constraints among the variables we use one of three methods:

(a) We use the method of Lagrangian multipliers, to incorporate the constraint. An example of this is provided by the kinematic fitting problem dealt with in detail in Section 5.2.2, and in problems 5.10 and 5.11.

(b) We change the variables to another (smaller) set that are independent. For example, if our unknowns are various fractions F_i satisfying the constraint

$$\Sigma F_i = 1, \tag{4.50}$$

it is simplest to define a large F_k in terms of all the other F_i by

$$F_k = 1 - \sum_{i \neq k} F_i. \tag{4.51}$$

Then these other F_i are regarded as the independent† variables of the problem.

---

† The F_i still have to satisfy $\Sigma_{i \neq k} F_i < 1$, but provided we have chosen a large F_k as the dependent fraction, this constraint is not likely to be violated, and the remaining F_i can be regarded as independent.

(c) We make use of an already existing computer program which will minimise a function subject to specified constraints.

Provided that the necessary dependent variables can be simply expressed in terms of the others, Method (b) is probably the easiest to use.

Problems

The problems for this chapter and the next one are to be found together at the end of Chapter 5.

5

Detailed examples of fitting procedures

In this chapter, we illustrate in greater detail some of the fitting procedures described in the previous one. Some further examples for the reader to attempt are to be found in the problems at the end of the chapter.

5.1 Least squares fitting

We start with some examples of determining parameters by the least squares method of Section 4.5.

5.1.1 Function linear in parameters – the straight line

We have already mentioned that it is particularly easy to find the values of the parameters when the fitted function is linear in them. The example we consider here is the very familiar one of fitting a straight line to a set of data points. Thus the fitted function is

$$y = a + bx, \tag{5.1}$$

where a and b are the unknown parameters. (N.B. This example is simple because it is linear in a and b, and *not* because y is linear in x.) The data consist of n points $(x_i, y_i \pm \sigma_i)$, i.e. we assume that the x-co-ordinates are exactly known, but that there is an uncertainty σ_i in the y-co-ordinate of each point (see Fig. 5.1). A physical example could be the determination of the expansion coefficient of a material by measuring its length (with an experimental uncertainty) at a series of well defined temperatures.

According to our prescription (4.40), we have to minimise

$$S = \sum_{i=1}^{n} \left(\frac{y_i - a - bx_i}{\sigma_i} \right)^2, \tag{5.2}$$

where the summation extends over all the n data points. Notice that in fact the deviation of each point from the line is defined as the difference in the

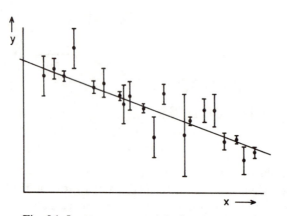

Fig. 5.1. Least squares straight line fit. The data consist of a series of points $(x_i, y_i \pm \sigma_i)$, whose x-co-ordinates are assumed to be known exactly, but whose y-co-ordinates have (varying) experimental uncertainties σ_i. The problem is to find that line such that the sum of the weighted squares of its deviations from all the points is smaller than that for any other possible line. The deviations are measured not as the shortest distance from each point to the straight line, but simply in the y direction. The weighting factor for any point is inversely proportional to the square of its error σ_i.

y-co-ordinate of the point and of the line at the particular x-co-ordinate of that point; it is not the closest distance that the line approaches the point. This is consistent with the philosophy that the x-co-ordinate is determined with no error. (Contrast Section 5.1.3, where we deal with the case of fitting the best straight line to a series of points with errors in both their x- and y-co-ordinates.)

On differentiating eqn (5.2) partially with respect to a and b respectively, we obtain

$$-\frac{1}{2}\frac{\partial S}{\partial a} = \sum \frac{y_i - a - bx_i}{\sigma_i^2} = 0 \tag{5.3}$$

and

$$-\frac{1}{2}\frac{\partial S}{\partial b} = \sum \frac{(y_i - a - bx_i)x_i}{\sigma_i^2} = 0. \tag{5.4}$$

The eqns (5.3) and (5.4) are two simultaneous equations for our two unknowns, which yield

$$b = \frac{[1][xy] - [x][y]}{[1][x^2] - [x][x]}, \tag{5.5}$$

where the quantities in square brackets are defined by

$$[f] = \frac{1}{n} \Sigma \frac{f_i}{\sigma_i^2}. \tag{5.6}$$

Note that the weighted means of the same quantities are given by

$$\langle f \rangle = [f]/[1]. \tag{5.7}$$

Then a is determined by rewriting eqn (5.3) as

$$\langle y \rangle = a + b \langle x \rangle. \tag{5.8}$$

Eqn (5.8) shows that the best line (in the least squares fitting sense) passes through the weighted centre of gravity of all the data points (with the weighting of each point as usual being inversely proportional to σ_i^2).

As we have repeatedly emphasised, now that we have obtained the best values of the parameters, we must go on and determine how accurately they are known. To calculate the errors, we must evaluate $\frac{1}{2}\partial^2 S/(\partial p_i \, \partial p_j)$, where p represents the vector of parameters. From (5.3) and (5.4), we obtain

$$\left.
\begin{aligned}
\frac{1}{2}\frac{\partial^2 S}{\partial a^2} &= n[1], \\[2mm]
\frac{1}{2}\frac{\partial^2 S}{\partial b^2} &= n[x^2] \\[2mm]
\text{and} \quad \frac{1}{2}\frac{\partial^2 S}{\partial a \, \partial b} &= n[x].
\end{aligned}
\right\} \tag{5.9}$$

Thus the inverse error matrix is

$$n \begin{pmatrix} [1] & [x] \\ [x] & [x^2] \end{pmatrix}, \tag{5.9'}$$

which we invert to obtain the error matrix for a and b as

$$\frac{1}{nD} \begin{pmatrix} [x^2] & -[x] \\ -[x] & [1] \end{pmatrix}, \tag{5.10}$$

where the determinant D is

$$D = [x^2][1] - [x][x]. \tag{5.11}$$

Thus we see that the calculation of the errors in this case is no more difficult than that of the actual quantities themselves!

Comment (i)

It is of the utmost importance that, before we try to fit a given set of data by the above procedure, we plot them on graph paper. This may indicate that one of the data points is in wild disagreement with the rest and should be ignored, that the data lie on a parabola rather than a straight line, that the data are linear only for a given range of x-values, etc. It also enables us to make some estimate of a and b, from which we can check that the values deduced from eqns (5.5) and (5.8) are sensible, which they may not be if we have made some mistakes in entering the data into the computer or calculator.

Comment (ii)

Our procedure is simply parameter determination, with no mention as yet of hypothesis testing. Thus we should insert the best values of a and b from (5.5) and (5.8) into eqn (5.2) in order to obtain S_{min}, and then look up the probability of getting a value as large as this in a χ^2-distribution with $n-2$ degrees of freedom (see Section 4.6). If this probability is unacceptably low, then our values of a and b are meaningless.

This test is essentially a quantitative expression of comment (i).

As a computational aside, it is worth noting that it is not necessary to loop over the data twice in order to obtain the value of S_{min} since a little algebra shows that

$$S_{min} = \Sigma \, y_i^2/\sigma_i^2 - a \, \Sigma \, y_i/\sigma_i^2 - b \, \Sigma \, x_i \, y_i/\sigma_i^2.$$

This is the 2-dimensional equivalent of the fact that the variance of a set of measurements can be calculated in a single loop over the data (see eqns (1.4) and (1.5), and the remarks following them).

Comment (iii)

From the error matrix (5.10) we see that

$$\mathrm{cov} \, (a, b) \sim -\langle x \rangle.$$

(Note that the quantities [1], n and D are necessarily positive.) Hence, if $\langle x \rangle = 0$ (and if it is not, then we can arrange for it to be so by a suitable change of the origin of the x-axis), then the covariance vanishes. This implies that the errors on the gradient of the line and on its y-value at the weighted centre of gravity of the points (through which, as we earlier demonstrated, the best fit line passes) are uncorrelated.

If, however, $\langle x \rangle$ is positive, then the covariance is negative. As can be

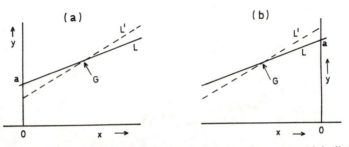

Fig. 5.2. An illustration of the fact that for a least squares straight line best fit, the covariance of the gradient b and the intercept a is proportional to $-\langle x \rangle$, the x-co-ordinate of the weighted centre of gravity of the data points (G in the diagrams). The best fit line L passes through G. If the gradient is increased by its error (to give the line L'), then the intercept a will decrease if $\langle x \rangle$ is positive (diagram (a)), or will increase if $\langle x \rangle$ is negative (diagram (b)).

seen from Fig. 5.2, this corresponds to the fact that if the gradient of the best fit line were chosen a bit *larger*, then the line pivots about the centre of gravity and the intercept on the y-axis *decreases*, i.e. the correlation between the gradient and the intercept is negative.

The covariance term can thus be regarded as expressing the fact that when $\langle x \rangle \neq 0$, there is an extra contribution to the error in the intercept arising from the uncertainty in the gradient of the best line through the weighted centre of gravity of the data points.

Comment (iv)

As expected, and as can be checked from (5.10), (a) all the terms of the error matrix have the correct dimensions, and (b) if all the errors σ_i are increased by a factor of F, then the terms of the error matrix all increase by the factor F^2.

Comment (v)

Another interesting property is obtained by considering the situation where $\langle x \rangle = 0$. (Again, if this is not already so, we transform the x-values to make it so.) Then from (5.8)

$$a = \langle y \rangle, \tag{5.12}$$

i.e. the intercept is simply the weighted average of all the y-values. If all the errors σ are equal, then from (5.10)

$$\sigma_a{}^2 = \frac{[x^2]}{n([x^2][1]-[x]^2)}$$

$$= \frac{1}{n[1]}$$

$$= \sigma^2/n. \tag{5.13}$$

Thus we obtain the eminently reasonable result that the error on the intercept is just $1/\sqrt{n}$ times the error on each of the measurements.

Comment (vi)

Thus far we have obtained the error matrix for the intercept and gradient of the fitted line. How well is the y-co-ordinate of the fitted line known for any particular x-value? We can regard eqn (5.1) as giving y as a function of a and b; since their error matrix is known we can use eqn (3.41) to obtain the variance on y as

$$\sigma_y{}^2 = x^2\sigma_b{}^2 + 2x\,\mathrm{cov}\,(a,b) + \sigma_a{}^2. \tag{5.14}$$

Again, if we transform our variables to ensure that $\langle x \rangle$ is zero, the covariance term vanishes and (5.14) simplifies to

$$\sigma_y{}^2 = x^2\sigma_b{}^2 + \sigma_a{}^2. \tag{5.14'}$$

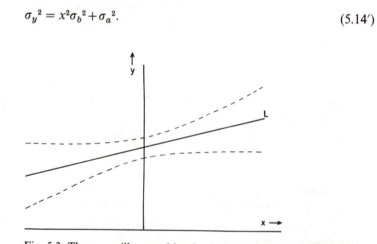

Fig. 5.3. The error, illustrated by the dashed curves, in a least squares best fit line L. The x-co-ordinates have been transformed to make $\langle x \rangle = 0$. Near the y-axis, the error in y is dominated by the uncertainty in the intercept and is constant. Far from the axis, the error is dominated by the uncertainty in the gradient, and is proportional to $|x|$.

Table 5.1. *Comparison of the relative merits of various methods of treating errors on data points in least squares fitting*

Method I uses the σ_i as given. Method II ignores them and estimates σ from the scatter of points about the fitted line. Method III compromises in using σ_i but scales them all to make $S_{min} = n - r$.

	Method I	Method II	Method III
Data points with big σ_i	Essentially ignored in fit	Treated like points with small σ_i	As for I
Errors on a and b	Realistic in terms of statistics of data (i.e. σ_i and n)	Can be unfortunately small if points happen to lie very well on straight line	As for II, unless we force $F \geqslant 1$
What if data really don't lie on straight line? (See Fig. 5.4)	Errors on a and b may be ridiculously small if statistics are high	Errors will be larger	As for II
No. of data points needed to estimate a, b and errors	2	3	3
Can goodness of fit be tested?	Yes	No	No
Can method be used if σ_i are unknown?	No	Yes	No

The interpretation of this is very simple. Far from the centre of gravity, the first term on the right-hand side dominates and $\sigma_y \sim \pm x\sigma_b$; the uncertainty in y is given by the uncertainty in gradient, and hence is proportional to the distance from the centre of gravity. But near the centre of gravity, the last term is the more important and $\sigma_y \sim \sigma_b$, i.e. the uncertainty is constant and simply equal to the error on the intercept (see Fig. 5.3).

Comment (vii)

We have calculated the error matrix (5.10) by making use of the given errors σ_i on the data points. There is another possibility (Method II of Table 5.1) of using the scatter of the individual points from the best fit line. To do this, we proceed as follows. We ignore the quoted errors† σ_i (to enable us to use the previous notation, we at this stage regard them all as set equal to unity) and recalculate the best fit line using eqns (5.5) and (5.8); this may differ somewhat from our previous best fit since we have changed the values of σ_i which enter into (5.5) and (5.8). Then S_{min} is obtained from (5.2), and is used to estimate the errors (assumed to be equal on all the points) by

$$\hat{\sigma}^2 = \frac{S_{min}}{n-r}, \tag{5.15}$$

where r is the number of free parameters (i.e. 2). Finally the error matrix can be obtained from (5.10), using the value of $\hat{\sigma}^2$ just obtained.

A compromise possibility (Method III) is to use the quoted errors σ_i but to scale them all by the same factor F which is chosen to make

$$S_{min} = n - r.$$

The relative merits of the three approaches are given in Table 5.1. As can be seen, Methods I and Methods III each have their own advantages in different circumstances as far as error calculations are concerned. A conservative approach is to estimate errors both ways and to take the larger answer; this is equivalent to using Method III with the constraint that

$$F \geqslant 1.$$

Finally we remark that the situation is similar to that of trying to find the average of two measurements which, for example, give answers of

---

† Of course, if for some reason the errors σ_i on the data points are not known, then this is the only way we can continue.

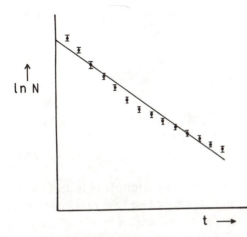

Fig. 5.4. The danger of estimating errors from (5.10). In high energy scattering experiments, the number of interactions N as a function of the momentum transfer t may be approximately exponential. Then $\ln N \sim a + bt$, but perhaps there is a small quadratic term in t. If we perform a high statistics experiment and we attempt to fit using just the linear form, then the parameters a and b will be determined with artificially small errors if we use Method I (i.e. using the experimental values of δN_i on each data point). Thus if we change slightly the range of t over which the fit is performed, the values of a and b can change by much more than their errors. If Method II or III is used (i.e. the errors are based on the scatter of points about the best fit line), the calculated errors will be larger.

1.0 ± 0.1 and 11.0 ± 0.2.† Then eqns (1.38) and (1.39) give us an answer of 3.0 ± 0.1; this corresponds to our Method I. Ignoring the errors, we obtain the average as 6 ± 7 (Method II). Finally we can scale up the quoted errors by a factor of ~ 45 in order to make the results consistent, as described for Method III; this gives an average of 3.0 ± 4.0.

5.1.2 *General least squares case*

Having dealt with the straight line case, we now consider the more general problem of trying to fit a set of data points $(x_i, y_i \pm \sigma_i)$ by a form $y^{th}(x, \alpha_a)$ which is a function of the variable x and involves the parameters α (of which there are a in number) in a way which is not necessarily linear.

† As in Comment (ii) of Section 1.6, we are using hypothetical numbers which no-one in his right senses would average, but they do serve the purpose of dramatising the way the various approaches produce different answers.

An example of this is the Breit–Wigner function (4.20), with parameters M_0 and Γ.

As before, we define

$$S = \sum_{i-1}^{n} \left(\frac{y_i - y^{th}(x_i, \alpha_a)}{\sigma_i} \right)^2.$$

(5.16)

The solutions for α are given by solving as simultaneous equations

$$\frac{\partial S}{\partial \alpha_a} = 0.$$

(5.17)

This is, of course, easier said than done. We return later in this section to a way of homing in on the solution provided we have reasonably good approximations for the α_a. Otherwise we minimise S by one of the other methods suggested in Section 4.7.

The errors on the parameters are given by

$$\langle (\alpha_a - \bar{\alpha}_a)(\alpha_b - \bar{\alpha}_b) \rangle = (H^{-1})_{ab},$$

(5.18)

where the matrix

$$H_{ab} = \frac{1}{2} \frac{\partial^2 S}{\partial \alpha_a \partial \alpha_b}.$$

(5.19)

Now let us consider the general linear case (i.e. where y^{th} is linear in the parameters α_a)

$$y^{th} = \sum_a \alpha_a f_a(x),$$

(5.20)

where the f_a are defined functions of x. Thus, for example, any polynomial involving positive and/or negative powers of x with arbitrary coefficients is in this sense linear. Then

$$S = \sum_j \left(\frac{y_j - \sum_a \alpha_a f_a(x_j)}{\sigma_j} \right)^2.$$

(5.21)

(N.B. We are using the subscript j for the data points, and a and b for the various parameters and their corresponding functions.) On differentiating, we obtain

$$-\frac{1}{2} \frac{\partial S}{\partial \alpha_b} = \sum_j \frac{(y_j - \sum_a \alpha_a f_a(x_j)) f_b(x_j)}{\sigma_j^2} = 0,$$

(5.22)

which are our b equations for our b unknown parameters α. Eqn (5.22) looks much neater in matrix notation:

$$Y_b - \sum_a \alpha_a H_{ab} = 0, \tag{5.22'}$$

where

$$Y_b = \sum_j \frac{y_j f_b(x_j)}{\sigma_j^2} \tag{5.23}$$

and

$$H_{ab} = \frac{1}{2} \frac{\partial^2 S}{\partial \alpha_a \partial \alpha_b}$$

$$= \sum_j \frac{f_a(x_j) f_b(x_j)}{\sigma_j^2}. \tag{5.24}$$

We can abbreviate (5.22') even more as

$$\mathbf{Y} = \alpha \mathbf{H}, \tag{5.22''}$$

which has as solution

$$\alpha = \mathbf{Y} \mathbf{H}^{-1}, \tag{5.25}$$

which most clearly demonstrates that the solutions to least square problems which are linear in the parameters can be directly obtained simply by matrix inversion, without our needing to become involved in the problems of minimisation procedures.

As in practical problems we will want to evaluate numerical values of α, we rewrite (5.25) in longhand as

$$\alpha_b = \sum_a \sum_j \frac{y_j f_a(x_j)}{\sigma_j^2} (H^{-1})_{ab}. \tag{5.25'}$$

Eqn (5.25') looks somewhat fearsome; the only way really to appreciate what it means is to use it to evaluate the best fit parameters in some actual problem. (Section 5.1.1 provides a particularly simple example of this.)

The error matrix for the parameters α is simply given by

$$\langle (\alpha_a - \bar{\alpha}_a)(\alpha_b - \bar{\alpha}_b) \rangle = H_{ab}^{-1}.$$

There is thus less work involved in determining the error matrix than in obtaining the parameters.

As we have the best fit parameters and their complete error matrix, we can calculate $y^{th}(x_j)$ and its error at each of the fitted points. Thus we can regard the fitting procedure for a set of data points as giving us corrected values of the y_j (lying on the fitted curve) and with reduced errors, i.e. the extra information that the data correspond to a function of known form improves the quality of the results.

Having seen how to deal with problems involving functions which are linear in the parameters we return to the non-linear case, but assume that we have an approximate solution β_a for the parameters which we wish to convert to the correct solution γ_a.

By definition, the correct values γ_a for the parameters are such that the derivatives $\partial S/\partial \alpha_a$, which are functions of the parameters $\alpha_1, \alpha_2, \alpha_3, ..., \alpha_b$, are all automatically zero. But if we use the approximate values for the parameters, then in general they will not be zero. We then correct the values β_a by assuming that we are near enough to the solution to have to consider variations of the derivatives which are only linear in the parameters. Thus we write

$$\frac{\partial S}{\partial \alpha_a}(\gamma_1 ... \gamma_b) = \frac{\partial S}{\partial \alpha_a}(\beta_1 ... \beta_b) + \sum_c \frac{\partial}{\partial \alpha_c}\frac{\partial S}{\partial \alpha_a}(\beta_1 ... \beta_b) \times (\gamma_c - \beta_c),$$

$$(5.26)$$

where we have explicitly written the fact that the derivatives are functions of the parameters. In eqns (5.26), the factors $(\gamma_c - \beta_c)$ are the corrections which must be used to get from the current values of the parameters to the correct values which satisfy eqn (5.17). The terms on the left-hand side are zero from the definition of the $\gamma_1 ... \gamma_b$. The important feature is that the corrections $(\gamma_c - \beta_c)$ occur linearly in the a simultaneous eqns (5.26), which we can thus invert to obtain the corrections. This is clearer if we rewrite (5.26) in matrix form as

$$\mathbf{D} = -\mathbf{E}\delta,$$

$$(5.26')$$

with solution

$$\delta = -\mathbf{E}^{-1}\mathbf{D},$$

$$(5.27)$$

where

$\delta_c = \gamma_c - \beta_c,$ the correction term for each parameter

$D_a = \dfrac{\partial S}{\partial \alpha_a},$ the vector of derivatives

and

$E_{ca} = \dfrac{\partial^2 S}{\partial \alpha_c \partial \alpha_a},$ the matrix of second derivatives.

Because we have omitted higher order terms in the expansion (5.26), the corrections will not exactly bring the β_c-values of the parameters up to their correct values γ_c. Thus it is probably necessary to iterate, by repeating as

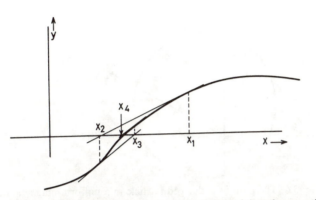

Fig. 5.5. The Newton–Raphson method for solving the equation $y(x) = 0$. For a point x_i sufficiently near the solution, we can obtain an improved estimate $x_{(i+1)} = x_i - y(x_i)/y'(x_i)$, which can be used iteratively to converge on the solution. The way the method works is illustrated in the diagram. The procedure described in the text for solving the simultaneous non-linear equations $\partial S/\partial \alpha = 0$ is the multi-dimensional equivalent of this.

many times as necessary the recipe (5.27), but with D recalculated with the values of the derivatives at the new estimates of the parameters. (Whether it is also necessary or desirable to recalculate the matrix of second derivatives E depends on such factors as how close to the original solution we were, how fast we want the process to converge, the amount of computation involved in recalculating E^{-1} at each stage, etc.)

The above procedure is simply the extension to several dimensions of the Newton–Raphson method of solving $y(x) = 0$ by using the values of $y(x_i)$ and $y'(x_i)$ at some point near the correct solution (see Fig. 5.5).

Once we are reasonably satisfied with the values of the parameters (i.e. all the derivatives are close to zero), the error matrix is as usual simply given by $(\tfrac{1}{2}E)^{-1}$.

5.1.3 Straight line fit to points with errors on x and y

We now turn to the problem of fitting a straight line to a set of n data points, on both of whose co-ordinates x_i and y_i there are uncertainties δx_i and δy_i.† Thus, for example we may be trying to measure the expansion coefficient of a metal by taking readings of the length and temperature of a bar as it is heated; both the length and temperature could have experimental uncertainties. Or we could be trying to fit a helix to the path

† For simplicity, we assume that the errors on x and y are uncorrelated.

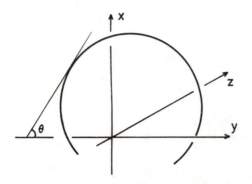

Fig. 5.6. The path of a charged particle in a uniform magnetic field, whose axis is along the z-direction. The particle describes a helix such that the tangent direction θ at any point is linearly related to the z-co-ordinate there.

of a charged particle in a magnetic field, where we have taken measurements of the particle's direction θ (projected onto the plane perpendicular to the field) and its distance z measured along the direction of the magnetic field; then θ and z are linearly related (see Fig. 5.6) but both are subject to experimental errors.

To find the straight line which provides the best fit to the data, we proceed more or less as in the case of the more standard example of Section 5.1.1 where the x-co-ordinates were assumed to be exactly determined, except for the way in which we specify by how far any line misses each point. In the standard case, we chose the difference in y-co-ordinates of the point and of the line for that x-co-ordinate, and divided by the y error. Then

$$S = \sum_{i=1}^{n} \left(\frac{y_i - a - bx_i}{\sigma_i} \right)^2. \tag{5.28}$$

For our case, however, we find for each data point that position on the straight line which minimises r_i, where

$$r_i = \frac{\text{distance of that position on line from given data point}}{\text{radius vector of error ellipse in that direction}}. \tag{5.29}$$

(See Fig. 5.7.) Thus r_i^{min} corresponds to that point on the line which minimises the error ellipse function

$$\frac{(x - x_i)^2}{\delta x_i^2} + \frac{(y - y_i)^2}{\delta y_i^2}. \tag{5.30}$$

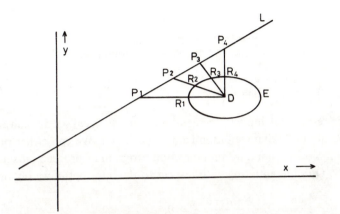

Fig. 5.7. The principle used in fitting a straight line to a set of data points D_i with errors on both their x- and y-co-ordinates, δx_i and δy_i. (Only one data point is shown in the diagram.) The errors define an error ellipse E. We can consider a series of point $P_1 P_2 P_3 P_4 \ldots$ on any given straight line L. For each of these, we evaluate the ratio of the distance DP_j to the radius DR_j of the error ellipse in that direction. For some particular point on the line, the ratio will have its smallest value r_i^{min}. (This corresponds to the height at the lowest point for the surface defined by the function $[(x-x_i)/\delta x_i]^2 + [(y-y_i)/\delta y_i]^2$, subject to the constraint of remaining on the straight line L.)

Then for any particular line, we construct

$$S = \sum_{i=1}^{n} (r_i^{min})^2. \tag{5.31}$$

Finally the best straight line is that which minimises S of eqn (5.31), with respect to variations in the gradient b and the intercept a.

After a little simple geometry, we derive r_i^{min} as

$$(r_i^{min})^2 = \frac{(y_i - a - bx_i)^2}{b^2 \delta x_i^2 + \delta y_i^2}. \tag{5.32}$$

This then leads us to another interpretation of r_i^{min}. It is simply the usual difference

$$d_i = y_i - a - bx_i$$

in the y-co-ordinates of the data point and of the line at that particular x_i value, divided by the error in d_i

$$\varepsilon_d^2 = \delta y_i^2 + b^2 \delta x_i^2, \tag{5.33}$$

where the last term in the above equation arises because, in this case, the *x*-co-ordinate is not precisely known.

Thus we are faced with the problem of minimising

$$S = \Sigma \frac{(y_i - a - bx_i)^2}{\delta y_i^2 + b^2 \delta x_i^2}. \tag{5.34}$$

It is the term involving b in the denominator which causes difficulties and in general, an analytic solution cannot be found. We can either put the function S into a standard minimisation program or proceed ourselves as follows. On differentiating with respect to the intercept a, we obtain

$$-\frac{1}{2} \frac{\partial S}{\partial a} = \Sigma_i \frac{(y_i - a - bx_i)}{\delta y_i^2 + b^2 \delta x_i^2} = 0. \tag{5.35}$$

Thus the intercept is given in terms of the gradient by

$$a = \frac{\Sigma \left(\dfrac{y_i - bx_i}{\delta y_i^2 + b^2 \delta x_i^2} \right)}{\Sigma \left(\dfrac{1}{\delta y_i^2 + b^2 \delta x_i^2} \right)} \tag{5.36}$$

and all that remains is to find the best value of b. This is most simply obtained by inserting a series of values of b (each with its corresponding value of a from eqn (5.36) into the expression for S (eqn (5.34)), and finding the one which gives the smallest value for S.

Now we must calculate the error matrix for a and b. We thus wish to know those values of a and b which increase S from its minimim value to $S_{min} + 1$. The method described below can also be used as another way of finding the best values of the two parameters.

We assume that in the neighbourhood of the minimum, S can be expressed in terms of the parameters a and b as

$$S - S_{min} = C_1(a - a_0)^2 + C_2(b - b_0)^2 + 2C_3(a - a_0)(b - b_0), \tag{5.37}$$

where a_0 and b_0 are those values of a and b which minimise S. Thus if S is evaluated at six different (a, b) values in the neighbourhood of the minimum, then the constants C_1, C_2, C_3 as well as a_0, b_0 and S_{min} can be calculated. A sensible configuration of points is as shown in Fig. 5.8; the algebra becomes particularly simple if

and
$$\left. \begin{array}{l} b_5 - b_2 = b_2 - b_4 \\ a_3 - a_2 = a_2 - a_1. \end{array} \right\} \tag{5.38}$$

The location of point 2 should be close to the minimum; and the step sizes

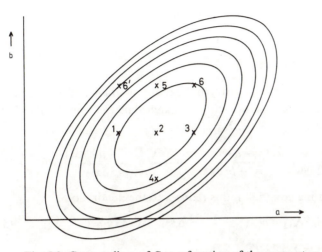

Fig. 5.8. Contour lines of S as a function of the parameters a and b. The points 1, ..., 6 are in a suitable configuration for determining the error matrix of the variables a and b. The function S is calculated at the six points, and then the six unknowns of eqn (5.37) are determined. The coefficients C_1, C_2 and C_3 are the elements of the inverse error matrix. Point 2 should be chosen close to the minimum M, and the grid sizes in the two directions should correspond to approximately one error. The location of point 6 is suitable when a and b are uncorrelated or if the correlation is positive (as shown in the figure). If the correlation is negative, then 6′ is a better choice for the last point.

in the two directions should be approximately equal to the respective errors. These can be well estimated from the first stage of the minimisation.

Finally the (a, b) error matrix is obtained by inverting the matrix

$$\begin{pmatrix} C_1 & C_3 \\ C_3 & C_2 \end{pmatrix}. \tag{5.39}$$

There is one case in which the algebra simplifies significantly and that is when the errors δx_i are the same for all the points, and similarly for δy_i. Then eqn (5.34) reduces to

$$S = \frac{\sum_i (y_i - a - bx_i)^2}{b^2 \delta x^2 + \delta y^2}. \tag{5.34'}$$

On differentiating, we obtain

$$-\frac{1}{2}\frac{\partial S}{\partial a} = \frac{\sum (y_i - a - bx_i)}{b^2 \delta x^2 + \delta y^2} = 0. \tag{5.35'}$$

Since the denominator is independent of i, we obtain

$$\Sigma \, y_i - na - b \, \Sigma \, x_i = 0$$

or

$$a = \langle y \rangle - b \langle x \rangle, \tag{5.40}$$

i.e. the best fit line passes through the centre of gravity of the data points, as in the standard case of fitting a straight line (see eqn (5.8)). We then differentiate eqn (5.34′) with respect to b,

$$-\frac{1}{2} \frac{\partial S}{\partial b} = \frac{\Sigma \, (y_i - a - bx_i) \, x_i}{\delta y^2 + b^2 \, \delta x^2} + \frac{b \, \delta x^2 \, \Sigma \, (y_i - a - bx_i)^2}{(\delta y^2 + b^2 \, \delta x^2)^2} = 0. \tag{5.41}$$

On substituting eqn (5.40), this reduces to the quadratic

$$b^2 \, \delta x^2 \, \Delta xy - b(\delta x^2 \, \Delta y^2 - \delta y^2 \, \Delta x^2) - \delta y^2 \, \Delta xy = 0, \tag{5.42}$$

where

$$\Delta x^2 = \Sigma \, x_i^2 - (\Sigma \, x_i)^2 / n,$$

and

$$\Delta y^2 = \Sigma \, y_i^2 - (\Sigma \, y_i)^2 / n$$

$$\Delta xy = \Sigma \, x_i \, y_i - \Sigma \, x_i \, \Sigma \, y_i / n.$$

In the standard solution of the quadratic, we choose the negative sign in front of the square root to give the best line; the positive sign gives the gradient of the worst fit straight line through the centre of gravity of the data points.

Comment (i)
This fitting procedure reduces to the standard one of Section 5.1.1 in the case where the δy_i are all set to zero.

Comment (ii)
This procedure is symmetric in the two variables, i.e. we obtain the same best fit straight line whether we fit in y against x, or in x against y.

5.2 Kinematic fitting

5.2.1 General comments

We now have an example of a least squares fitting procedure which incorporates constraints among the variables by the use of Lagrangian multipliers. The problem is that of the kinematic fitting of an observed interaction by a particular hypothesis for the reaction, and was briefly mentioned in Section 4.6.2.

We again assume that we have observed and measured a four prong interaction (as in Fig. 4.13) and we wish to test whether this is an example of the reaction

$$pp \rightarrow pp\pi^+\pi^-.$$

The way this is done in principle is to consider all configurations of momentum vectors of the outgoing particles which conserve energy and momentum. Then from this infinity of sets of possible momentum vectors we choose the one which is closest (in the least squares sense) to the observed set of vectors.

For the case where the errors on the measured variables are uncorrelated, we thus construct†

$$S = \sum_{tracks} \sum_{x,y,z} \left(\frac{p_i^{meas} - p_i^{fit}}{\sigma_{p_i}}\right)^2, \tag{5.43}$$

where the measured momenta are $p^{meas} \pm \sigma_p$, and p^{fit} are the possible fitted values. If the measured variables are correlated, we define

$$S = \sum_{i=1}^{3N} \sum_{j=1}^{3N} (p_i^{meas} - p_i^{fit}) E_{ij}(p_j^{meas} - p_j^{fit}), \tag{5.43'}$$

where the sum extends over the three components of momentum p_i for each of the N tracks, and E_{ij} is their inverse error matrix. Then we calculate S_{min} by varying the p^{fit}, subject to the constraints of energy and momentum conservation; it is at this stage that the Lagrangian multipliers enter the problem.

The reason for using this elaborate procedure is two-fold.

Reason (i)

We can convert the value of S_{min} into a probability that our hypothesis about the reaction is incorrect, which then enables us to decide whether or not to accept this interaction as an example of the $pp\pi^+\pi^-$ final state. We thus look up the actual value of S_{min} in χ^2 tables, where the relevant number of degrees of freedom is simply the useful number of energy and momentum constraints. (Examples of this are given in more detail in Appendix 2.)

† In practice we use $1/p$, $\tan\lambda$ and θ as variables, rather than $p_x p_y$ and p_z.

Reason (ii)

We obtain improved estimates of the track variables, i.e. the extra information of energy and momentum conservation enables us to obtain the particles' momenta with an accuracy better than that from the measurements alone (cf. the second paragraph below eqn (5.25′) in Section 5.1.2). Furthermore, the fitted momenta are physically consistent in that they satisfy the conservation equations.

Reasons (i) and (ii) above correspond to the familiar problems of hypothesis testing and parameter determination respectively.

In real life, we have to test several hypotheses for each measured interaction (corresponding to different particle assignments for the seen charged tracks and for possible unseen neutrals). Hopefully only one of these yields a satisfactory probability; we then think we know what the particle assignments are and the interaction is said to be 'unique'. Sometimes, however, at least two hypotheses give a satisfactory probability and the interaction is said to be 'ambiguous'. It is a dangerous approach in ambiguous cases simply to accept as correct the hypothesis with the largest probability. It is impossible to give a general procedure for dealing with such ambiguities. Each set of interactions that are ambiguous between a particular pair of hypotheses must be investigated in detail before criteria can be defined for assigning them sensibly between the possible hypotheses. In general this will result in not every interaction being correctly assigned; it is thus important to have an idea of both the loss of correct events and the contamination of incorrect ones (errors of the first and second kind respectively) in the finally accepted sample for a given reaction's hypothesis.

It is also worth remembering that in deciding whether or not to accept a particular hypothesis concerning a given event, we often have extra information other than just the momentum vector for a track. Thus, for example, it may have passed through Cerenkov counters or have measurable ionisation (either of which gives information concerning the particle's velocity), it may be observed to decay or interact in some characteristic manner, etc.

5.2.2 Actual fit in a simplified case

Kinematic fitting is an involved process, and large and complicated computer programs exist for fitting interactions in very general configurations. In order to illustrate the basic principles involved, however, we will treat a very simplified situation.

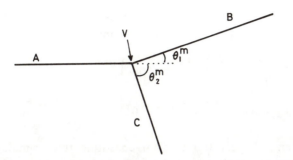

Fig. 5.9. An observed 2-prong interaction. The beam track A interacts at the point V with a proton to produce two outgoing tracks B and C whose production angles are measured as $\theta_1{}^m \pm \varepsilon$ and $\theta_2{}^m \pm \varepsilon$ respectively, but whose momenta are unknown. We wish to test whether this is an example of the reaction pp → pp. If this is so, then in the non-relativistic limit, we expect the production angles to add up to 90°.

We assume that we have observed an interaction involving two outgoing charged tracks (see Fig. 5.9). The apparatus involves no magnetic field, and hence the magnitudes of the particles' momenta are unknown. The angles $\theta_1{}^m$ and $\theta_2{}^m$ are measured with equal accuracy ε, and are assumed independent.

We wish to test whether our observed interaction is an example of the elastic scattering reaction

$$p + p \to p + p. \tag{5.44}$$

Then, assuming that the energies of the particles are small enough for us to use non-relativistic kinematics, we must satisfy the constraint†

$$C = \theta_1 + \theta_2 - \frac{\pi}{2} = 0, \tag{5.45}$$

where θ_1 and θ_2 are the true values of the tracks' angles, which we are trying to obtain.

Then as usual we define

$$S = \left(\frac{\theta_1{}^m - \theta_1}{\varepsilon}\right)^2 + \left(\frac{\theta_2{}^m - \theta_2}{\varepsilon}\right)^2. \tag{5.46}$$

We have to minimise S with respect to the variables θ_1 and θ_2, subject to the constraint C. Since we are pretending that this is an example of a

† We should also check that the incident and outgoing track directions are coplanar, but we are ignoring this too.

realistic 'minimisation subject to kinematic constraints' type of problem, we proceed to solve it by using a Lagrangian multiplier λ:

$$\left.\begin{array}{l} \dfrac{\partial S}{\partial \theta_1} + \lambda \dfrac{\partial C}{\partial \theta_1} = -2\dfrac{(\theta_1{}^m - \theta_1)}{\varepsilon^2} + \lambda = 0, \\[4mm] \dfrac{\partial S}{\partial \theta_2} + \lambda \dfrac{\partial C}{\partial \theta_2} = -2\dfrac{(\theta_2{}^m - \theta_2)}{\varepsilon^2} + \lambda = 0. \end{array}\right\} \tag{5.47}$$

Then (5.47) and the constraint (5.45) constitute three simultaneous equations for the three unknowns λ, θ_1 and θ_2. On solving them, we obtain

and

$$\left.\begin{array}{l} \theta_1 = \theta_1{}^m + \dfrac{1}{2}\left(\dfrac{\pi}{2} - \theta_1{}^m - \theta_2{}^m\right) \\[4mm] \theta_2 = \theta_2{}^m + \dfrac{1}{2}\left(\dfrac{\pi}{2} - \theta_1{}^m - \theta_2{}^m\right). \end{array}\right\} \tag{5.48}$$

Since θ_1 and θ_2 are known functions of $\theta_1{}^m$ and $\theta_2{}^m$, we can use the transformation (3.46) to obtain the error matrix of θ_1 and θ_2. In this case, however, the relations (5.48) are so simple that we can almost immediately deduce that

$$\delta\theta_1 = \delta\theta_2 = \varepsilon/\sqrt{2} \tag{5.49}$$

and that $\delta\theta_1$ and $\delta\theta_2$ are completely anti-correlated, as expected (since θ_1 and θ_2 must add up to $\pi/2$).

Finally we insert (5.48) into (5.46) to obtain

$$S_{min} = \frac{\left(\dfrac{\pi}{2} - \theta_1{}^m - \theta_2{}^m\right)^2}{2\varepsilon^2}. \tag{5.50}$$

This should be distributed as χ^2 with one degree of freedom (corresponding to our one constraint eqn (5.45)) and enables us to decide whether or not to accept our hypothesis (5.44) for this particular interaction.

Comment (i)

The test function (5.50) for our hypothesis looks sensible, since the fit is acceptable if

$$\theta_1{}^m + \theta_2{}^m \approx \frac{\pi}{2}$$

and the error on the left-hand side $\sqrt{2}\,\varepsilon$.

Non-linearities could arise via the constraint equations, or if the function g is more complicated than a quadratic in x and y. In that case it is simplest to recast the equations into a form suitable for iterative solution; the variables x, y and z are recalculated at each step in terms of their values at the end of the previous iteration.

After this mathematical aside, we now return to our original problem and rewrite eqn (5.43′) in terms of the corrections a_i to the i measured variables that are required in order to satisfy the constraints, i.e.

$$C_j(m_i + a_i, u_k) = 0.$$

where u_k are the required values of the unmeasured variables. Then

$$S = \sum_i \sum_l G_{il} a_i a_l, \qquad (5.43'')$$

G_{il} being the elements of the inverse error matrix.

The derivatives of this expression with respect to the corrections a_i are

$$\frac{\partial S}{\partial a_i} = 2 \sum G_{il} a_l. \qquad (5.59)$$

Thus the equivalent of eqns (5.56) and (5.57) becomes (in matrix notation)

$$GA + D^T \Lambda = 0, \qquad (5.60)$$

while that of (5.58) can be written

$$E^T \Lambda = 0, \qquad (5.61)$$

where

G is the inverse error matrix of size $i \times i$,

Λ is the vector of the j Lagrange multipliers,

D and E are the derivatives of the constraint equations with respect to the measured and unmeasured variables, and are of dimension $j \times i$ and $j \times k$ respectively,

A is the vector of the corrections to the measured variables.

In general the constraint equations corresponding to (5.54) and (5.55) e non-linear in the variables, but provided we are not too far from the rrect solution, we may expand them by Taylor series. Keeping only the ms linear in the deviations from the best values, we obtain

$$DA + E\Delta U + R = 0, \qquad (5.62)$$

$$R = C(M + A°, U°) - DA°,$$

C is the vector of the j constraint imbalances that is obtained when using the current estimates of the unmeasured variables ($U°$) and

Comment (ii)

Our final values θ_1 and θ_2 as given by eqns (5.48) of course satisfy the constraint (5.45). This is desirable since it avoids ambiguities in calculating kinematic variables for the interaction. In contrast, if we use the *measured* quantities, we must make a choice. Thus for example, if we want the production angle of the first track, we can use either $\theta_1{}^m$ or else $\pi/2 - \theta_2{}^m$, and these are not exactly equal.

Comment (iii)

As expected, the accuracy (5.49) of the final fitted values is better than that of the measured quantities. The improvement is a factor of $\sqrt{2}$. This is understandable since $\theta_1{}^m$ and $\pi/2 - \theta_2{}^m$ can be regarded as two independent estimates of θ_1, each of accuracy ε. Hence our final answer for θ_1 will be their average (which is just what eqn (5.48) says), whose variance is down by a factor of two. (See eqns (1.38) and (1.39).)

Comment (iv)

We found the exact solution (5.48) for θ_1 and θ_2 after a very simple calculation. This was because our problem was linear in the variables. In general this is not so, and kinematic fitting becomes an iterative procedure.

5.2.3 Fit involving unmeasured variables

In the example of the previous section, we assumed that all the required variables that occurred in the constraint equations were measured. We now deal with a more general case where one or more of these quantities may be initially unknown. Examples of this could be

(i) $\Lambda \to p\pi^-$,

where the direction of the neutral Λ is known, but its momentum is not;

(ii) $pp \to pp\pi^+\pi^-$,

where one of the outgoing particles is of unknown momentum; or

(iii) $pp \to np\pi^+\pi^+\pi^-$,

where the neutron's momentum and directions are undetermined (i.e. three missing quantities).

As in the previous example, the fitting procedure is reduced to the mathematical problem of minimising S of eqn (5.43′) with respect to the measured variables, subject to satisfying the constraint equations, which now contain both the measured and the unmeasured quantities. (Thus in Example (i) above, the momentum and energy constraints all involve the unmeasured magnitude of the momentum of the $\Lambda°$.)

We denote the original values of the measured quantities by m_i and their final fitted values by f_i; the final values of the unmeasured quantities are u_k; and the constraints $C_j = 0$ are satisfied exactly by the variables f_i and u_k. We also use i, j and k to denote the numbers of measured variables, constraints and unmeasured variables respectively.

In order for us to be able to determine the unknowns u_k from the constraint equations, we must have $j \geqslant k$. With $j = k$, the u_k can in general be calculated without altering the measured quantities m_i. Thus the final solution will have $S = 0$, but the number of degrees of freedom is $j - k = 0$, so this is not a fitting problem. Thus interesting cases will all have

$$k < j. \tag{5.51}$$

Furthermore if the total number of measured plus unmeasured variables $i + k$ is less than the number of constraints, it will in general be impossible to choose values of the $i + k$ variables which satisfy all j constraint equations simultaneously. With $i + k = j$, the variables can be calculated exactly from the constraint equations, and the relevant sum of squares S *calculated* directly. This is a physically meaningful situation, but to see how *fitting* procedures operate, we require

$$i + k > j. \tag{5.52}$$

In our problem of $\Lambda°$ decay (see Fig. 5.10), $i = 8$, $j = 4$ and $k = 1$ so both (5.51) and (5.52) are satisfied. The simplest non-zero values of i, j and k giving an interesting example for understanding the procedure are 2, 2 and 1 respectively.

A purely mathematical analogy is thus the minimisation of a function, say

$$g(x, y) = a^2(x - x_0)^2 + b^2(y - y_0)^2, \tag{5.53}$$

subject to some constraints, say

$$C_1(x, y, z) = x + y + z = 0 \tag{5.54}$$

and

$$C_2(x, y, z) = x - y - 2z - 1 = 0. \tag{5.55}$$

The variables x and y in g play the role of the measured variables appearing in the sum S to be minimised, while the extra variable z appearing only in the constraint equations is equivalent to an unmeasured variable. The problem is solved by introducing the Lagrange multipliers λ_1 and λ_2 for

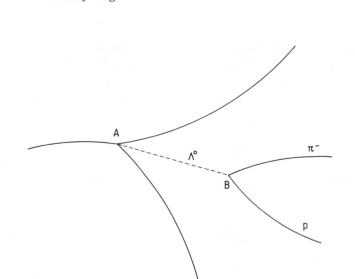

Fig. 5.10. Appearance of tracks in a visual detector in a magnetic fi[eld]. The incident particle enters from the left and interacts at A, produ[cing] two charged tracks and a neutral one (which would be unseen in [the] detector, but is shown as a dashed line in the diagram). At B, the neutral particle decays to two charged ones. We are interested i[n] testing the hypothesis that the measured tracks labelled π^- and [p are] consistent with being the decay products of a $\Lambda°$, whose mome[ntum is] unknown in magnitude, but which was travelling along the di[rection] AB. For the hypothesis $\Lambda \rightarrow p\pi^-$, we have eight measured qu[antities] (the 3-momenta of the proton and of the π^-, and the 2-dire[ction] cosine ratios of the $\Lambda°$); one unmeasured quantity (the ma[gnitude of] the $\Lambda°$'s momentum); and four constraints (energy and 3-[momentum] conservation at the $\Lambda°$ decay point).

the two constraint eqns (5.54) and (5.55). Then

$$\frac{1}{2}\frac{\partial g}{\partial x} + \lambda_1 \frac{\partial C_1}{\partial x} + \lambda_2 \frac{\partial C_2}{\partial x} = 0,$$

$$\frac{1}{2}\frac{\partial g}{\partial y} + \lambda_1 \frac{\partial C_1}{\partial y} + \lambda_2 \frac{\partial C_2}{\partial y} = 0,$$

$$\lambda_1 \frac{\partial C_1}{\partial z} + \lambda_2 \frac{\partial C_2}{\partial z} = 0.$$

Then the five eqns (5.54–5.58) are sufficient to fi[nd] z and the two Lagrange multipliers.

In the example given above, the five equation[s] and the Lagrangian multipliers, and so their so[lution]

the corrections (A°) to the measured ones,

and

ΔU is the vector of corrections to the current estimates U° of the unmeasured variables.

This then forms the basis of an iterative procedure for calculating the values of the i measured and k unmeasured variables that satisfy the constraint equations (and of the j Lagrange multipliers). The eqns (5.60–5.62) are linear in A, ΔU and Λ and we can thus calculate the corrections to the measured and unmeasured variables.

The conventional method of solving the eqns (5.60–5.62) is to use considerable matrix manipulation to obtain

$$\left.\begin{aligned} \Delta U &= -(E^T G_D E)^{-1} E^T G_D R, \\ \Lambda &= G_D(E\Delta U + R) \\ A &= -G^{-1}D^T \Lambda, \end{aligned}\right\} \tag{5.63}$$

and

where

$$G_D^{-1} = DG^{-1}D^T$$

and is the error matrix for the vector DM. With this procedure, none of the matrices involved is larger in size than the larger of $(i \times i)$ or $(j \times j)$.

In view of the large number of matrix multiplications and the corresponding possibility of programming errors, it is not necessarily disadvantageous to rewrite the set of linear eqns (5.60–5.62) as

$$My = z,$$

where M is the square matrix of size $i+j+k$ given by

$$\begin{pmatrix} G & 0 & D^T \\ 0 & 0 & E^T \\ D & E & 0 \end{pmatrix},$$

z is the vector

$$\begin{pmatrix} 0 \\ 0 \\ -R \end{pmatrix}$$

and y is the vector containing the corrections to the measured and unmeasured variables

$$y = \begin{pmatrix} A \\ \Delta U \\ \Lambda \end{pmatrix} = M^{-1}z. \tag{5.64}$$

Thus only one matrix operation is required per iteration, and this is the method we recommend.

In either of these approaches, the improved, fitted estimates of the variables are approximately linear functions of the original measured values, and hence their error matrix can be obtained by trivial but tedious algebra.

Before the fit proper begins, it is necessary to obtain starting values for the unmeasured variables, by using the measured variables and some of the constraint equations. These calculated unmeasured variables together with the measured variables then form the zeroth iteration. From them we calculate the constraint unbalances C and the numerical values of D and E, the derivatives of the constraints with respect to the measured and unmeasured variables. Then the matrix eqn (5.64) is evaluated in order to obtain the corrections to the measured and unmeasured variables, which can then be used as starting values for the next iteration.

It is useful to check before each iteration that the values of the variables are within their allowed range. For example, direction cosine ratios must certainly be of magnitude smaller than unity, and the physical configuration of the apparatus may well impose a smaller limit. In general, genuine events do not transgress these limits during the iteration procedure.

For an actual example of the reaction we are testing, the zeroth iteration values for the variables will not exactly satisfy the constraint equations; for a random combination of tracks, the constraint unbalances will in general be large. As the iteration proceeds, the constraint unbalances are reduced and the new estimates of the measured variables move away from their original measured values. This continues until the constraint imbalances are reduced to a preset level, at which stage the final value of S is calculated from equation (5.43″). Events with S satisfactorily small are accepted as being examples of the tested reaction; a limit in the region of 10 would be satisfactory for a fit involving three effective constraints.

5.3 Maximum likelihood determination of Breit–Wigner parameters

We now turn to an example of determining parameters by the maximum likelihood method. We assume we are provided with an experimentally determined set of mass values m_i, which we wish to fit by a Breit–Wigner resonance function together with some innocuous background. We may, for example, be interested in determining the best estimates of M_0 and Γ for a recently discovered resonant state; or we may want to determine how

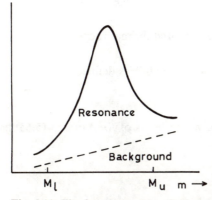

Fig. 5.11. The function used to fit the mass spectrum, consisting of a resonance plus background. The data are a set of mass values, rather than simply a histogram, and are not shown. The fit is performed to the n observed mass values in our chosen mass range M_l to M_u. As well as helping to obtain the best values of the parameters, the computer should tell us what fraction F_1 of the events between M_l and M_u is due to the resonance, and what fraction F_2 of the resonance events lies in the tails beyond our accepted mass range. Then the total number of resonance events is $nF_1/(1 - F_2)$.

much resonance of known M_0 and Γ there is, in order to calculate the production cross-section for the resonance. We choose for the form of the fitted function

$$f(m) = B\text{–}W(m) + Bgd(m), \tag{5.65}$$

where

$$B\text{–}W(m) = \frac{\Gamma/2}{(m - M_0)^2 + (\Gamma/2)^2} \tag{5.66}$$

and

$$Bgd(m) = a\{1 + b(m - \overline{M}) + c(m - \overline{M})^2\}, \tag{5.67}$$

where M_0 and Γ are the parameters of the Breit–Wigner function (respectively the mass and width of the resonance), while a, b and c parametrise the background (describing the amount of background, its slope and curvature respectively) (see Fig. 5.11). The fitting procedure is carried out for masses over the range M_l to M_u,† whose mean is $\overline{M}$. Note that, simply for ease of integration (see eqn (5.70) below) in this demonstration of the method, we have chosen to add the Breit–Wigner

---

† Subjective judgement is involved yet again in choosing this range (compare the remarks at the end of Section 2.3).

function to the background (physics may suggest that it is better to choose a form $Bgd \times (1 + B - W)$) and that we have taken a simple form for our Breit–Wigner function (we have ignored the mass dependence of the width, etc.).

The logarithm of the likelihood function is defined by

$$\ell = \sum_{\text{events}} \log y_i, \qquad (5.68)$$

where for the ith event, y is given in terms of $f(m_i)$ of eqn (5.65) as

$$y_i = \frac{f(m_i)}{\displaystyle\int_{M_l}^{M_u} f(m) \, dm}. \qquad (5.69)$$

The integral of f can be obtained analytically.†

$$\int f(m) \, dm = \tan^{-1}\left(\frac{M_u - M_0}{\Gamma/2}\right) + \tan^{-1}\left(\frac{M_0 - M_l}{\Gamma/2}\right)$$
$$+ a(M_u - M_l)\left(1 + \frac{c}{12}(M_u - M_l)^2\right). \qquad (5.70)$$

To find the best values of the parameters, we must vary them so that they maximise ℓ. If there are only a few parameters (although five is rather large for this) to be varied, then we can probably enjoy ourselves by looking for the maximum in our own computer program. We could get the computer to vary the parameters one at a time, to fit a parabola through those values near the maximum, and to tell us its estimate of the best value and error for that parameter. Thus, for example:

We tell the computer:–

> Select events with masses in the range 0.74–0.82 GeV. Keep the following parameters fixed
> $M_0 = 0.785$ GeV,
> $\Gamma = 0.019$ GeV,
> $b = c = 0$ (i.e. background is independent of mass).
> Vary a over the range 0–10.

Then the computer tells us:

> Best value of $a = 5.2 \pm 1.7$, corresponding to

---

† For more complicated forms of f (for example, if $B–W$ is replaced by a Gaussian function, or if $B–W$ contains a mass dependent width), then it is necessary to perform the integration numerically. Analytic integration should be performed if possible since it saves computer time, and the normalisation has to be recalculated at each stage of the maximisation.

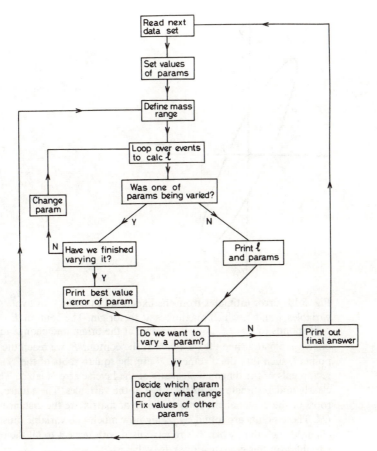

Fig. 5.12. Schematic flow diagram of a do-it-yourself computer programme to find the maximum of a likelihood function of a few variables, such as is used to fit a resonance plus background to a set of mass values.

86±4% of the events in the given mass range being in the resonance; and 17.7% of the resonance's tails lie outside the mass range.

Now we have to work round the other parameters, varying them. Then it will probably be necessary to go back to the first variable, and repeat the whole procedure several times (until we are convinced that we understand any possible correlations and are close to the maximum, have run out of computer time, or are just fed up). A schematic flow diagram of this procedure is shown in Fig. 5.12.

We have, of course, still to find the errors on the parameters. After we

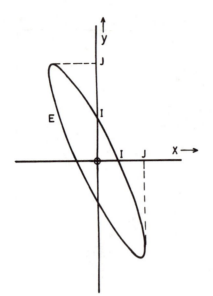

Fig. 5.13. Error estimates from the likelihood function ℓ of two variables x and y, which maximises at the origin. The contour E corresponds to $\ell = \ell_{max} - \frac{1}{2}$. If we start at the origin and change just one variable at a time till we arrive at the contour E, we reach the points I, such that the distances OI are the square roots of the reciprocals of the diagonal elements of the inverse error matrix. They clearly underestimate the likely range of the variables. The square roots of the diagonal elements of the error matrix are the distances OJ. These points are found by seeing how much one variable must be changed such that, when ℓ is remaximised with respect to all the other variables (in this case, just one), its value is $\ell_{max} - \frac{1}{2}$.

have found the maximum, we can vary the parameters one at a time to find out by how much they must change in order to make

$$\ell = \ell_{max} - \tfrac{1}{2}. \tag{5.71}$$

But this provides us with only the diagonal elements of the *inverse error* matrix. To obtain a diagonal element of the *error* matrix, we must find out by how much we must vary the corresponding parameter so that, when ℓ is remaximised with respect to all the other parameters, the condition (5.71) is satisfied. The situation for two parameters is illustrated in Fig. 5.13 (see also Section 3.4).

Although it is much more fun and instructive to do the maximisation ourselves as indicated above, it clearly becomes very cumbersome with more than a few variables. In particular, convergence may be very slow, and the calculation of the errors is tiresome. It thus becomes desirable to use a proper maximisation program (see Section 4.7) to find the solution for us.

Problems

5.1   An experiment observes the following ten values of $\cos\theta$ for the decay of a resonant state:

0.05, 0.15, 0.25, 0.35, 0.45, 0.55, 0.65, 0.75, 0.85, 0.95,

i.e. only ten examples of this state have been observed, with each of the above $\cos\theta$ values being obtained once. Use (a) the method of moments, and (b) the maximum likelihood method to fit the expression $N(1+b/a\cos^2\theta)$ to the data in order to estimate b/a and its error (N is a normalisation factor). Do you think that the function provides a satisfactory description of the data?

5.2   We wish to perform a fit by the maximum likelihood method of a Gaussian distribution (of variable mean μ and width σ) to a set of measurements x_i. By considering the relevant second derivatives, show that the variances of the estimates of μ and σ are σ^2/n and $\sigma^2/2n$ respectively, and that they are uncorrelated. Show also that if the fitted variables are chosen alternatively as μ and σ^2, the variance on the latter is $2\sigma^4/n$. Finally, investigate whether the variances on the estimates of σ and for σ^2 are related to each other as expected.

5.3 (i)   The decay rate λ is to be determined for a radioactive species, whose probability of decay at time t is proportional to $e^{-\lambda t}$. A total of n decays are observed, with the decays occurring at times $t_1, t_2, ..., t_n$.

Find a suitable normalisation constant for the decay expression, and then use the maximum likelihood method to show that the best estimate

$$\lambda_1 = n/\Sigma\, t_i,$$

i.e. it is given by the inverse of the mean of the observed decay times. Show also that the accuracy of this estimate is $\lambda_1/\sqrt{n}$.

(ii)   In the above situation, the efficiency for observing decays varies with time like $e^{-\nu t}$, where ν is a constant, i.e. it is unity for small t, but falls off at larger times. Use the formula (4.31) to show that the estimate λ_a is given by

$$\lambda_a = \Sigma \, e^{vt_i}/\Sigma \, t_i \, e^{vt_i},$$

i.e. λ_a is the reciprocal of the weighted mean of the times, where each time is weighted by the inverse of the efficiency corresponding to that time.

(iii) Use eqn (4.31') of the text, applied to the situation described in (ii) above, in order to deduce that

$$v + \lambda_b = n/\Sigma \, t_i,$$

i.e. if λ is estimated from the reciprocal of the mean of the observed unweighted times, it must be corrected by subtracting v, because of the experimental bias towards small observed times.

5.4   Perform a least square best fit of a straight line $y = mx + c$ to the following data

x	0	1	2	3	4
y	5 ± 0.1	7 ± 0.2	11 ± 0.1	12 ± 0.2	14 ± 0.1

(i) Calculate the intercept c and the slope m, and their error matrix. How probable is the fit?

(ii) Repeat (i), except in the data multiply all the errors on the y-measurements by a factor of 10.

(iii) Repeat the calculation (i), except ignore the quoted errors on the y-measurements, and use the 'deviation from the fit' method of calculating errors.

(iv) How probable is the fit of the line $y = 2x + 5$ to the data (for the case where the errors are multiplied by a factor of 10)?

5.5   Two variables x and y are related by the formula $y = a + bx$. The values of y and of the gradient dy/dx are measured at three different x-points; they are $y_i \pm \sigma_i$ and $(dy/dx)_i \pm \varepsilon_i$ respectively (for $i = 1, 2, 3$). Use a least squares method to obtain from these data the best values of a and b, and also their error matrix.

5.6   The least squares method is used to fit a straight line to a set of measurements $y_i \pm \sigma_i$ at various x_i values. Show that

$$\Sigma \left(\frac{y_i - y^{th}(x_i)}{\sigma_i} \right)^2 = \Sigma \, (y_i^2/\sigma_i^2) - b \, \Sigma \, (x_i \, y_i/\sigma_i^2) - a \, \Sigma \, (y_i/\sigma_i^2),$$

where $y^{th}(x_i)$ is the predicted y-value for the best straight line at the relevant x_i-points, and a and b are the best estimates of the intercept and the gradient respectively.

5.7 (a) Derive the formulae (1.38) and (1.39) for combining the results of different experiments which are attempting to measure the same quantity, as follows. Assume that the various experiments are consistent with a common value $\hat{a}$, and construct an appropriate sum of squares S to test this hypothesis. Minimise S with respect to $\hat{a}$ to obtain the best estimate of a, and use the second derivative of S at the minimum in order to calculate the error on a. Alternatively calculate the error on a directly from eqn (1.38), in terms of the errors on the individual measurements a_i.

(b) An experiment estimates two quantities r and θ as r_1 and θ_1 with an accuracy as given by the error matrix $\begin{pmatrix} \delta r_1^2 & \delta r_1\,\delta\theta_1 \\ \delta r_1\,\delta\theta_1 & \delta\theta_1^2 \end{pmatrix}$. A second experiment provides another determination r_2 and θ_2 with its error matrix which is the same as the previous error matrix but with the subscripts 1 replaced by 2. What is our best estimate for r and θ, and what is its error matrix? Verify that in a suitable limit, these answers are consistent with those of (a).

5.8 Two experiments attempt to measure the same quantity, and obtain the results 0.9 ± 0.1 and 1.4 ± 0.2. Decide whether these are consistent

(i) by looking up probabilities in the tails of Gaussian distributions, for an appropriate variable; and

(ii) by using a least squares method to test the hypothesis that the two measurements are consistent.

5.9 We have n measured points $(x_i, y_i \pm \sigma_i)$ where each of the y_i are assumed to be Gaussian distributed. Show that in fitting a function $y = y(x)$ which involves parameters p_k, the maximum likelihood and the least squares methods to determine p_k are equivalent. Show also that the recipes for determining errors that ℓ decreases by $\frac{1}{2}$ and that S increases by 1 give the same answers.

5.10 Estimates $f_i \pm \sigma_i$ are made of the fractions F_i of various subsamples of a population. It is believed that these subsamples are exclusive and complete, i.e. $\Sigma\, F_i = 1$. Use the method of Lagrangian multipliers to derive improved estimates of the fractions F_i by minimising $\Sigma (F_i - f_i)^2/\sigma_i^2$, subject to the constraint that $\Sigma F_i = 1$.

Use the above method to deduce the best estimates of the fractions in a 2 subsample case, where the measured fractions are 0.80 ± 0.05 and 0.1 ± 0.1. What is the accuracy of these estimates? At what level of probability are the data consistent with the hypothesis $\Sigma F_i = 1$?

5.11   Find the values of x, y and z which minimise the function (5.53) of the text, subject to the constraints (5.54) and (5.55).

5.12   A population has a Gaussian distribution with mean zero and unit variance. An estimate s^2 of the variance of the population is obtained from a sample of size N (larger than 1) by the formula

$$s^2 = \frac{1}{N-1} \sum_i^N (x_i - \bar{x})^2,$$

where x_i are the individual observations, and $\bar{x}$ is their mean. Given that $(N-1)s^2$ has a χ^2-distribution with $N-1$ degrees of freedom, and using the properties (4.45) and (4.45′) of χ^2 distributions, show that s^2 is an unbiassed estimate of the variance, and that its variance (i.e. $\overline{(s^2 - \sigma^2)^2}$) is $2/(N-1)$.

Find the expectation values and variances of other estimates

$$s_k^2 = \frac{1}{N+k} \sum_i^N (x_i - \bar{x})^2,$$

where k is a constant, and hence show that, although s_1^2 is biassed, its mean square deviation from unity (i.e. from the correct value of the variance) is smaller than for any other estimate s_k^2.

6

Monte Carlo calculations

6.1 Introduction

6.1.1 What is it?

The Monte Carlo approach basically provides a method of solving probability theory problems in situations where the necessary integrals are too difficult to perform. We thus start by showing how the Monte Carlo technique may be used to evaluate an integral

$$I = \int_a^b y(x)\,dx, \tag{6.1}$$

where $y(x)$ is any specified function (see Fig. 6.1).

A non-Monte Carlo method of evaluating (6.1) numerically is to divide the range from a to b into n equal steps, and then to use as an estimate of I the sum

$$\frac{b-a}{n} \sum_{i=1}^{n} y(x_i), \tag{6.2}$$

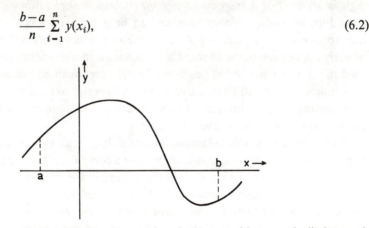

Fig. 6.1. The function $y(x)$ is to be integrated between the limits a and b. One method consists of finding the average value of y at suitable x-co-ordinates in this range, and then multiplying by $(b-a)$. The x-co-ordinates can be uniformly spread, or in the Monte Carlo method they are randomly distributed.

where
$$x_i = a + (i - \tfrac{1}{2})(b-a)/n. \tag{6.3}$$

(NB We are not suggesting that this is the best way of estimating the integral I numerically, but it bears the closest resemblance to the simple Monte Carlo approach.)

The corresponding Monte Carlo method consists of evaluating the sum (6.2), but instead of calculating the $y(x_i)$ at equally spaced intervals, we instead evaluate it at the randomly chosen points

$$x_i = a + r_i(b-a). \tag{6.4}$$

Here the r_i are members of a series of random numbers uniformly distributed in the range 0–1.

6.1.2 Random numbers

Random numbers have a specified distribution (which for the time being we take as being uniform in the range 0–1), but the individual values are unpredictable.

Sequences of random numbers can be obtained by drawing numbers out of a hat, from tables of random numbers (which can be found in most textbooks on statistics, or in statistical tables), from many pocket calculators or from the number n of decays observed in a fixed time interval from a radioactive source. In the last example, if n is odd or even, a single binary digit is set as 0 or 1 respectively; by m successive applications of this procedure an m-digit binary number can be generated, which is then transformed into the range 0–1. Another method, which was employed in an early application of the Monte Carlo technique in high energy physics, used the stopping azimuthal position of a cylinder which had been spun by a motor, which had been activated by an operator and turned off by a cosmic ray passing through a nearby detector. Yet another possibility utilises the electrical noise level in a resistor.

Most Monte Carlo calculations require a large number of random numbers, so it is sensible to obtain them from a computer. Such sequences are not random in the true sense, since they are generated by some algebraic algorithm, and hence are predictable. However, as far as Monte Carlo calculations are concerned, they have many of the properties of true random number sequences. They are called 'pseudo-random'.

Especially in using a computer-generated pseudo-random number sequence for the first time it is advisable to perform several checks.

   (i) Ensure that the random numbers are in the range 0 to 1, rather than -1 to $+1$, 0 to 100, or anything else. If not, the range can be changed by a suitable linear transformation.

  (ii) The numbers should indeed be (pseudo) random. Thus, at very least, they should be approximately uniformly distributed over their range, and successive pairs of numbers should be statistically uncorrelated.

 (iii) Because the sequence is only pseudo-random, it will eventually get back to where it started; this is the most important way in which pseudo-random series differ from genuinely random ones. If any particular problem requires the use of so many random numbers that this situation arises, then the whole basis underlying Monte Carlo calculations is lost and there is the distinct danger that the answer will be incorrect. It is thus prudent to check, for example, whether the first random number is repeated again in the sequence of random numbers used, since most generators would in this circumstance then proceed to repeat the subsequent random numbers.

Pseudo-random number sequences can be started either from a specified or from a random position in the sequence. The former is useful in testing programs, in that subsequent runs can then repeat the same calculations identically, and hence enable possible bugs to be investigated. The latter option is useful at the production stage of a program, so that consecutive runs can be combined to produce a statistically improved result. An even better approach for production running is to print out the initial and final random numbers for each run; then a subsequent run can be started just after the finishing position of the previous one.

6.1.3 Why do Monte Carlo calculations?

If we use n random numbers to obtain an estimate of the integral I of (6.1), the accuracy of the answer is proportional to $1/\sqrt{n}$, whereas that for even the simplest numerical method is proportional to $1/n^2$. What then is the point of doing Monte Carlo calculations?

If we were concerned with integrals over only one variable, then we would rarely use the Monte Carlo technique. Real life, however, often entails integration (either explicitly or implicitly) over a many dimensional region. In that case, the accuracy of the simplest numerical method is reduced to $n^{-2/d}$ (where d is the number of dimensions) , and more sophisticated numerical methods suffer a similar loss of accuracy; the error on the Monte Carlo calculation, however, remains at $n^{-\frac{1}{2}}$.

Another very important reason is that, in many cases, the boundary of the multi-dimensional region over which the integration is to be performed may be so complicated that it is in practice impossible to set up the required network of points at which the integrand is to be evaluated for the numerical method. In contrast, in the Monte Carlo method, it is merely necessary to test whether the randomly chosen point is inside or outside the region of integration; then the point is rejected, or the integrand evaluated respectively.

Another feature of Monte Carlo calculations is that, if we repeat the calculation, then we will get a slightly different answer each time (provided that we remember to start at a different place in the random number sequence on each occasion). Thus Monte Carlo calculations simulate real life situations, where the repetition of a particular experiment of limited experimental accuracy or of finite statistics is liable to produce a somewhat different result from the previous measurement.

6.1.4 Accuracy of Monte Carlo calculations

As with all statistical calculations, it is necessary in this case as well to provide an estimate of the accuracy of the calculation of the integral I.

The method that comes most readily to mind is to repeat the calculation several times, to use the spread of answers to calculate the variance of the distribution of estimates, and then to find the accuracy of their mean. (See Section 1.4.2.) This of course requires more computing time, so an alternative is to regard the initial calculation as consisting of a group of separate estimates; thus if n random numbers are used altogether, ten intermediate answers can be constructed, each utilising $n/10$ random numbers. The accuracy of the final answer is calculated as in the previous example.

The ultimate in dividing up the n random numbers is to have n estimates, each obtained from only one point. The width of the distribution of estimates will be simply σ, where σ^2 is the variance of the distribution of y-values. The final estimate of I will thus have an error of $\sigma/\sqrt{n}$.

Monte Carlo calculations often use very large amounts of computer time, since σ (especially in a multi-dimensional case) may be large compared with the average value of the integrand, and because the accuracy of the answer improves but slowly with n. Just as there are numerical methods of calculating the integral of (6.1) which are significantly better than the use of eqns (6.2) and (6.3), while not involving much more computation, so there are various techniques for obtaining a more accurate answer than provided by the crude Monte Carlo method of eqns (6.2) and (6.4). If serious use of Monte Carlo calculations is envisaged, it is important to incorporate such variance-reducing techniques were possible, as they can result in a considerable saving of computer time and/or an answer of improved accuracy.

Some of the techniques are:

(a) Stratification

The simple technique of dividing the range of integration into two equal x-regions, and having half the Monte Carlo generated points in each, reduces the variance. The reason is that it helps ensure a more uniform distribution of x-values than does the crude Monte Carlo method.

(b) Non-uniform sampling

Instead of generating points uniformly in x, it is advantageous to have a higher density where the integrand is varying more rapidly. The simplest step in this direction is to use a variant of method (a), in which the fractions of Monte Carlo points assigned to the two halves of the x-range are unequal, but are so chosen as to reduce the variance of the final answer. The calculation of the optimum fractions involves being able to perform our integration, but a reasonable approximate choice can often be made.

(c) Importance sampling

Since the variance of the Monte Carlo answer is proportional to that of the integrand, it is advantageous to transform the integral so that the new integrand has less variation than the original one. Thus we can write

$$\int y(x)\,dx = \int [y(x)/w(x)]\,w(x)\,dx$$

$$= \int [y(x)/w(x)]\,dv(x), \tag{6.5}$$

where
$$v(x) = \int w(x)\,dx.$$

The variance of the answer now depends on that of y/w, rather than that of y itself. By choosing the shape of w to resemble that of y, this variance can be made small. (Selecting w as being identical to y reduces the variance of y/w to zero. But being able to produce a random, uniform distribution in v requires us to be able to integrate w (or equivalently y) analytically. If we can do this, the use of a Monte Carlo approach is unnecessary.)

(d) Antithetic variates

For each Monte Carlo generated point x, we create according to some specified procedure another point x', and we use the contribution $y(x) + y(x')$ to the integral. With x' suitably chosen, $y(x)$ and $y(x')$ can be negatively correlated such that the variance on $y(x) + y(x')$ (and hence on the final answer) is reduced.

An example is that, for y varying monotonically with x over the range 0–1, we can choose $x' = 1 - x$. This is equivalent to symmetrising the function about the line $x = \frac{1}{2}$, and hence any component of y which is an odd function of $(x - \frac{1}{2})$ is removed from the integral.

(e) Adaptive techniques

These aim to learn about how to improve the variance of the answer as the integration proceeds. Thus we could imagine a trial search of the simple non-uniform sampling method with a smallish number of Monte Carlo points for various choices of the division between the two parts of the integration region, in order to enable an approximation to the best choice to be made for a larger calculation.

(f) Division and rejection

These are techniques which can be applied to phenomena involving sequences of processes, each of which often has a low probability of occurring. The division technique is used when the early stages of the sequence are successful; then that configuration is copied a fixed number of times, and each of them is then followed through the subsequent stages of the process.

The rejection method is employed in situations where the probability of each process in the chain is calculated separately, and a decision on whether to keep the whole configuration is made only at the end of the sequence on the basis of the overall probability, which is the product of the individual ones. If this product falls below a certain value at any stage, that sequence is aborted there and then, without bothering to follow it through to the end, and a new sequence is started from the beginning of the chain. This saves time by avoiding calculating the latter stages of combinations that are going to have very small probabilities of being accepted.

These techniques could be useful in, for example, following the development of an electromagnetic shower produced when a high energy electron is incident upon a thick detector.

6.1.5 More than one dimension

In order to integrate over n dimensions, we need n random numbers to generate the point at which the integrand is evaluated. Most regions of integration do not have limits which are independent of each other. It is then important to avoid the trap of unwittingly choosing one's random points such that their distribution is not uniform over the region of integration. Thus in order to evaluate $\int z(x, y)\mathrm{d}x\mathrm{d}y$ for the 2-dimensional triangular region of integration shown in Fig. 6.2, it would be incorrect to choose x randomly in the range 0–1, then y in the range 0–x and simply add up all $z(x, y)$-values. This procedure would incorrectly result in a higher density of points being obtained near the origin. Instead both x and y should be chosen in the range 0–1, and if the resulting point is such that $y > x$, it is thrown away and new values of *both* x and y are chosen.

The value of the integral can be calculated as either the analytic value of the area of the allowed region in the x–y plane multiplied by the average z for the Monte Carlo points within this area; or as the area of the rectangular region over which all the Monte Carlo x–y values are distributed, multiplied by the average z for all these points (where $z(x, y)$ is taken as zero for points outside the allowed region).

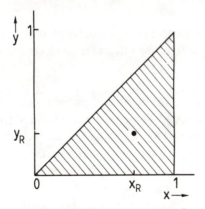

Fig. 6.2. In order to evaluate $\int z(x, y)\,\mathrm{d}x\,\mathrm{d}y$ over the shaded region, it is *incorrect* to add up the contributions $z(x_R, y_R)$ at points which are obtained by choosing x_R as a random point in the range 0–1, and then y_R randomly in the range 0 to $y(x_R) = x_R$.

Alternatively the original method of choosing $y < x$ can be used, but then allowance must be made in evaluating the summation for the non-uniform density of chosen points.

6.2 An integral by Monte Carlo methods

In order to illustrate the Monte Carlo technique in more detail, we use it to evaluate a very simple integral. This also enables us to see the extent to which the various variance reduction techniques are effective.

The integral we choose is

$$\int_0^1 x^3 \, dx.$$

For each of the methods, we use 128 randomly generated x points to evaluate the integral. We then repeat this procedure a further 39 times in order to see how much the result varies, and hence to measure the accuracy of a single estimate. The methods we employ are as follows:

(i) The crude Monte Carlo method, using eqns (6.2) and (6.4).

(ii) The 'hit and miss' Monte Carlo technique. After generating each x value, we decide whether to keep that point with a probability proportional to the y-value of the curve at that x. This is done by generating another random number r, and keeping the point provided that

$r < y(x)/[y(x)]_{max}$,

where $y(x)_{max}$ is the maximum value of the function in the range of interest. Then the estimate of the integral is given by

$I = (b - a) \times y(x)_{max} \times n_1/n,$

where n_1 and n are respectively the number of accepted points and the total number of points tested. (Thus whereas in method (i) we add $y(x)$ for each point, here we effectively add 1 with a probability proportional to $y(x)$.) This method thus determines the area by finding the probability that a point with randomly chosen x- and y-co-ordinates lies below the curve.

(iii) We divide the x range into two, and estimate the two sub-integrals separately using an equal number of Monte Carlo points in each range. The case where the division is chosen at $x = 0.5$ can be regarded as the simplest example of stratification, while comparison of the accuracies for different positions of the cut is an elementary example of an adaptive technique.

(iv) We divide the x-range into halves, but use unequal fractions of the random points in each half. Together with (iii), this can be regarded as a specific case of using a variable fraction f of the Monte Carlo points in a range up to a variable x_c, and the remaining $1-f$ in the range x_c to 1. In fact the optimum choice for our problem is $f \sim \frac{1}{2}$ and $x_c \sim \frac{2}{3}$.

(v) For each random x value, we also use the antithetic value $1-x$. Since this in fact uses a total of 256 x-values, we also perform the calculation using only half the original number of points.

(vi) We mimic importance sampling by choosing $w(x) = x^2$ in eqn (6.5). The integral then becomes

$$\int x^3 \, dx = \int (x^3/x^2) \, x^2 \, dx$$

$$= \frac{1}{3} \int x \, dx^3.$$

We thus generate events uniformly in the variable

$$v = x^3$$

over the v range 0–1, extract x, and find

$$\frac{1}{128} \Sigma \frac{x_i}{3}.$$

The reason for choosing $w(x) = x^2$ is that this is a function with a shape somewhat similar to the given $y(x) = x^3$. Of course using $w(x) = x^3$ would have been the obvious and best choice, reducing the error on the answer to zero. (Compare the remarks concerning importance sampling below eqn (6.5).)

The accuracies σ found for these different methods, together with the mean of the 40 determinations in each case, are presented in Table 6.1. As expected all the answers are within $2\sigma/\sqrt{40}$ of the correct answer of 0.25. We further see that the 'hit and miss' method is even less accurate than the crude Monte Carlo (as well as using twice as many random numbers). Stratification at $x = 0.5$ reduces the error, but choosing the cut at $x = 0.65$ is better still. In order to achieve the same accuracy the 'hit and miss' method would require about 30 times as many randomly generated numbers as the stratification method with

Table 6.1. *Evaluation of* $\int_0^1 x^3 \, dx$ *with 128 Monte Carlo points*

Monte Carlo method	Accuracy of single answer	Mean of 40 determinations
Crude Monte Carlo	±0.025	0.257
Hit and miss	±0.042	0.263
Stratify at $x = 0.5$	±0.015	0.255
Stratify at $x = 0.65$	±0.011	0.253
15% of points below $x = 0.5$	±0.013	0.253
Antithetic variates	±0.010	0.252
Antithetic variates with 64 points	±0.015	0.251
Importance sampling	±0.006	0.251

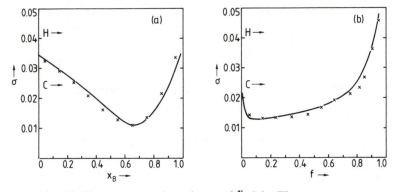

Fig. 6.3. The error σ on the estimate of $\int_0^1 x^3 \, dx$. The crosses are calculated using 128 Monte Carlo points when (a) they are divided equally in the two subranges 0 to x_B and x_B to 1; and (b) they are divided so that a fraction f is in the subrange 0 to 0.5, and the remaining $(1-f)$ in 0.5 to 1. The solid curves are the expected errors, as calculated from the variances of the y-values of the integrand in the two subranges. A small systematic shift of the crosses relative to their appropriate curves can arise because of our particular method of performing the Monte Carlo calculation (i.e. the *same* random numbers are used for each separate value of x_B (or of f)). At the extreme values of x_B, the curve in (a) tends to a value of $\sqrt{2}$ times the crude Monte Carlo error. The horizontal arrows marked C and H show the errors for the 'crude' and the 'hit and miss' Monte Carlo methods respectively. It is seen that either method of dividing the data can produce a significant reduction in the error.

the cut at $x = 0.65$. The antithetic variate technique also achieves a significant improvement in accuracy, even when using only half the number of randomly generated x-values. Our examples of importance sampling and non-uniform sampling (see Fig. 6.3) also result in a greatly reduced error.

It is important to remember that the improvement that will be achieved by each of these methods is dependent on the function to be integrated; used without thought in other cases it is possible to obtain a worse answer rather than a better one.

6.3 Very simple applications of Monte Carlo calculations

Having discussed several methods of performing integrals, we now go on to show how Monte Carlo techniques can be applied in very simple problems, one taken from mathematics and the other from high energy physics. These examples are in fact so simple that other techniques can be more usefully employed for their solution; we use them simply to illustrate the Monte Carlo technique.

6.3.1 Evaluation of π

If a very poor player throws darts at a board on which a circle is inscribed in a square, the ratio of darts falling in the circle to those within the square would be equal to the ratio of their areas, i.e. $\pi/4$. We can thus evaluate π by using pairs of random numbers to define the x- and y-co-ordinates of a point in the region

$$0 < x < 1, \quad 0 < y < 1.$$

Then the fraction of points which lie within the inscribed circle (i.e. which satisfy $(x-\frac{1}{2})^2 + (y-\frac{1}{2})^2 < \frac{1}{4}$)† provides our estimate $\pi/4$ (see Fig. 6.4(a)).

The expected accuracy of this method for determining π is $1.64/\sqrt{n}$, where n is the number of random number pairs used. There are, of course, non-Monte Carlo methods which evaluate π by series expansions, and which converge much faster than our example. There are also other Monte Carlo methods of calculating π (e.g. the Buffon's needle method – see Problem 6.2).

---

† The same answer can be achieved somewhat faster by checking whether $x^2 + y^2 < 1$, i.e. whether the chosen point lies within the top right hand quadrant of a unit circle centred on the origin (see Fig. 6.4(b)).

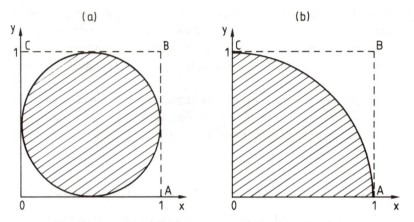

Fig. 6.4. A Monte Carlo method for evaluating π. Two random numbers are used to define a point's x- and y-co-ordinates in the range 0–1; such points uniformly cover the square $OABC$. We then check what fraction of the points lie within a circle of radius $\frac{1}{2}$ centred on $(\frac{1}{2}, \frac{1}{2})$. This provides us with an estimate of $\pi/4$. Alternatively we can find the fraction within the quadrant of the circle shown in (b).

6.3.2 Mean transverse momentum

In high energy physics, it is observed that the mean transverse momentum of particles produced in interactions rises only slowly as the incident energy increases. This has lead to many studies of transverse momentum p_T. One particular plot of interest is of the mean transverse momentum, as a function of the particles' centre of mass longitudinal momentum p_L. This often exhibits a dip for small values of p_L. Is this a sign of interesting physics, or does it have a more mundane explanation?

We can set ourselves the problem of deriving the $\langle p_T \rangle$ against p_L plot for the following assumptions. Particles are produced with centre of mass momenta p uniform in the range 0–1 GeV/c, and are also uniform in the cosine of the centre of mass production angle, $\cos \theta$. Neither of these assumptions is realistic, but it will serve to define a problem that we will solve by a Monte Carlo technique.

The Monte Carlo approach to this problem consists in generating a suitable large number (say 10000) tracks whose momentum is given by our usual random number, while the production angle is defined in terms of a second random number r by

$$\cos \theta = 2r - 1.$$

Table 6.2. *Program for* $\langle p_T \rangle$ *as* $f(p_L)$

```
C               GENERATE EVENTS UNIFORMLY IN P AND COS(TH)
C               ( IN RANGES 0 TO 1 AND -1 TO +1 RESPECTIVELY )
C               PLOTS AVERAGE PT V PL.
C               PT IS TRANSVERSE MOMENTUM
C               PL IS LONGITUDINAL MOMENTUM
        DIMENSION AVPT(21),ERPT(21),N(21)
        IC=123456
C               IC DEFINES START OF RANDOM NO. SEQUENCE
        NEV=10000
C               NEV IS NUMBER OF EVENTS TO BE GENERATED
C               NEXT ZERO LOCATIONS USED FOR SUMMATIONS
        DO 2 K=1,21
        AVPT(K)=0.
        ERPT(K)=0.
      2 N(K)=0
        DO 1 J=1,NEV
C               NEXT WE USE THE FUNCTION RAN TO PRODUCE RANDOM NUMBERS
C               IT PRODUCES A NEW RANDOM NUMBER EACH TIME IT IS CALLED
        PCM=RAN(IC)
        COSTH=2.*RAN(IC)-1.
C               PCM AND COSTH ARE C.M. MOMENTUM AND COS OF PRODN ANGLE
        PL=PCM*COSTH
        SINTH=SQRT(1.-COSTH*COSTH)
        PT=PCM*SINTH
        IBIN=20.*ABS(PL)+1.
C               EVENTS ARE BINNED ACCORDING TO THEIR PL
C               IBIN DEFINES BIN FOR PL OF THIS EVENT
        IF(IBIN.LE.20.AND.IBIN.GE.1)GO TO 3
        PRINT 100,PCM,COSTH,PL,PT,IBIN
    100 FORMAT(' WHY IS IBIN OUT OF RANGE?',4E13.3,I5)
        IBIN=21
C               THIS IS THE OVERFLOW AND UNDERFLOW BIN
      3 CONTINUE
        CALL AV(PT,IBIN,AVPT,ERPT,N)
C               SUBROUTINE AV CALCULATES FOR EACH BIN OF PL
C                  AVPT = SUM OF PT
C                  ERPT = SUM OF PT*PT
C                  N    = NUMBER OF EVENTS
C               FOR ALL EVENTS IN THAT BIN
      1 CONTINUE
        CALL FINAV(AVPT,ERPT,N)
C               SUBROUTINE FINAV CONVERTS AVPT, ERPT AND N INTO
C                  AVPT = AVERAGE PT
C                  ERPT = ERROR ON AVPT
C               FOR EACH BIN OF PL
        DO 4 K=1,21
      4 PRINT 101,K,N(K),AVPT(K),ERPT(K)
    101 FORMAT(' BIN',I3,' HAS',I4,' POINTS WITH AVPT=',F6.3,
      1        ' +-',F6.3)
        END
```

With p_L and p_T calculated as

$$p_L = p \cos \theta$$

and

$$p_T = p \sin \theta,$$

we obtain the average p_T for the set of events with p_L in specified ranges. The program to do this is reproduced in Table 6.2.

The average values of p_T are plotted as a function of the p_L-range in Fig. 6.5. It is seen that indeed there is a significant dip at low p_L. This arises from the fact that the tracks with small p are forced to have both p_T and p_L small, and it is this which introduces the correlation between the two

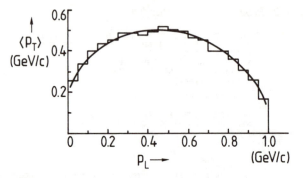

Fig. 6.5. The average value $\langle p_T \rangle$ of the transverse momentum plotted as a function of p_L for tracks produced isotropically in space and with momenta uniformly distributed over the range 0–1 GeV/c. The histogram is the result of a Monte Carlo calculation with 10 000 tracks, while the curve is an analytic calculation. Because of the expected symmetry about $p_L = 0$, the histogram and graph are shown only for positive p_L.

variables. The fall-off at large p_L simply comes from the fact that $p_T^2 + p_L^2 \leqslant 1$, and hence p_T is again constrained to be small.

Again with this problem, it is not necessary to use the Monte Carlo technique to find the solution. Analytic integration gives

$$\langle p_T \rangle = \frac{\sqrt{(1-p_L^2)} - p_L \tan^{-1}(\sqrt{(1/p_L^2 - 1)})}{-\ln p_L}.$$

This is drawn as the full curve in Fig. 6.5. The evaluation of this by computer requires less time than the Monte Carlo calculation, but its derivation takes longer, and the possibility of getting the wrong answer is larger.

6.4 Non-uniform distributions and correlated variables

So far we have used the random number generator to provide us with a uniform distribution of x-co-ordinates in the integration problem of 6.2, or of momenta in the example of Section 6.3.2. In many cases, however, we are interested in producing a non-uniform distribution $y(x)$ where, of course, $y(x) \geqslant 0$ for all physical values of x. For example, a mass distribution may be of the Breit–Wigner type

$$\frac{dn}{dm} = y(m) = \frac{\Gamma/2}{(m-M_0)^2 + (\Gamma/2)^2}. \tag{6.6}$$

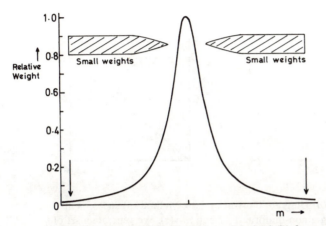

Fig. 6.6. The weighting technique (Method (i)) for producing a non-uniform distribution $y(m)$. Thus to generate a Breit–Wigner distribution we choose masses m_i uniformly distributed over a suitable range spanning the peak (for example, between the arrows). Then each mass is weighted by a different factor $y(m_i)$. When the distribution is required for a range larger than Γ, many of the m_i (corresponding to the shaded regions) will have small weights relative to those in the peak of the distribution, and these events are effectively thrown away.

Alternatively, the number of events as a function of a variable z $(0 \leqslant z \leqslant 1)$ may be linear

$$\frac{\mathrm{d}n}{\mathrm{d}z} = y(z) = 2z. \tag{6.7}$$

There are three standard methods for producing such non-uniform distributions.

Method (i)
We generate events uniformly in the variable x, but then weight each event by the appropriate factor $y(x)$. Then each Monte Carlo event is characterised by its variable x_i and its weight $y(x_i)$. (See Fig. 6.6.) This is equivalent to the 'crude' Monte Carlo method used for integration.

Method (ii)
We again generate events uniformly in the variable x, but will keep the event only if

$$\frac{y(x_i)}{y_{max}} > r_i, \tag{6.8}$$

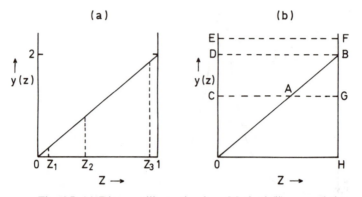

Fig. 6.7. (*a*) Diagram illustrating how Method (ii) succeeds in producing a non-uniform distribution $y(z)$. We generate a set of random numbers $z_1, z_2, z_3, \ldots$ uniformly in the range 0–1, and for each we calculate $f_i = y(z_i)/y_{max}$. In this case $y(z)$ is $2z$, and y_{max} is two. Then for each z, we generate another random number r_i, and keep the corresponding z_i only if $f_i > r_i$. Thus for z_1, f_1 is small and so this z-value is very likely to be rejected. For $z_3, f_3 \approx 1$ and so this value is almost certain to be kept. For an intermediate point $z_2, f_2 \approx 0.4$, and so there is a 40% probability that r_2 will be smaller than f_2 and that this z-value will be retained. Thus the distribution of the events which are retained approximates to $y(z)$. This method is equivalent to using pairs of random numbers to generate points uniformly in the y–z plane, and then keeping the corresponding z-value only if the point lies below the desired curve. (*b*) Method (ii) depends on knowing the value of y_{max} (the line DB). If we overestimate y_{max} (for example, the line EF), then we simply reject more events than is necessary, since f_i is then always smaller than unity. In this case, the fraction of rejected events is given by the ratio of the areas $OEFB$ to $OEFH$ (rather than $\frac{1}{2}$ for the correct choice of y_{max}). If y_{max} is underestimated (e.g. the line CG), then $f_i > 1$ for all values of z greater than that for the point A, and hence all such z-values are automatically retained. Thus the distribution of the kept events is distorted to OAG, rather than following the correct form OAB.

where y_{max} is the maximum value of $y(x)$ over the whole physical range of x, and r_i is yet another random number which is calculated afresh for each event. For distributions (6.6) and (6.7) the values of y_{max} are $2/\Gamma$ and 2 respectively. A schematic explanation of how this method works in the latter case is shown in Fig. 6.7(a). This method is equivalent to the 'hit and miss' Monte Carlo integration procedure.

Method (iii)

If we have a uniform distribution in some variable β, but then transform our variable to $x = x(\beta)$, the distribution in x will in general no longer be uniform. Thus to produce a specified non-uniform distribution $y(x)$, our problem is to find the correct transformation. Now the distributions of points in x and in β are related by

$$\frac{\mathrm{d}n}{\mathrm{d}x} = \frac{\mathrm{d}n}{\mathrm{d}\beta}\frac{\mathrm{d}\beta}{\mathrm{d}x},$$

and as the distribution in β is uniform (i.e. $\mathrm{d}n/\mathrm{d}\beta = 1$), then

$$\frac{\mathrm{d}\beta}{\mathrm{d}x} = \frac{\mathrm{d}n}{\mathrm{d}x} = y(x),$$

whence the required transformation is given by

$$\beta(x) = \int^{x} y(x)\,\mathrm{d}x. \tag{6.9}$$

Thus, for the Breit–Wigner (6.6), we write

$$y(m)\,\mathrm{d}m = \mathrm{d}\left[\tan^{-1}\left(\frac{m - M_0}{\Gamma/2}\right)\right]$$

$$= \mathrm{d}\beta, \tag{6.10}$$

i.e. we generate events uniformly in β, over the range $-\pi/2$ to $+\pi/2$, and then obtain m from eqn (6.10) as

$$m = M_0 + \frac{\Gamma}{2}\tan\beta. \tag{6.10$'$}$$

Then m will automatically be distributed according to the Breit–Wigner form.

Alternatively for the distribution (6.7), we have

$$y(x)\,\mathrm{d}x = \mathrm{d}(x^2)$$

$$= \mathrm{d}\beta. \tag{6.11}$$

Once again, we generate events uniformly in β (over the range 0–1), but then calculate the variable of interest x by

$$x = \sqrt{\beta}, \tag{6.11$'$}$$

which will then have the distribution (6.7).

A graphic explanation of how this method succeeds in producing the

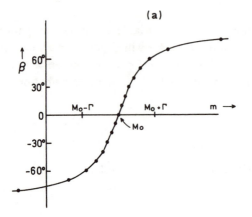

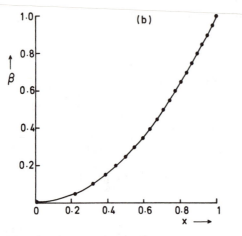

Fig. 6.8. Diagrams illustrating Method (iii) of producing non-uniform distributions. By the choice of a suitable function $\beta(m)$ or $\beta(x)$, a uniform distribution in β is used to generate the required distribution in m or x. The circles along the curves are equally spaced in β; their corresponding m- or x-values give the desired distribution. (In an actual Monte Carlo calculation, the distribution in β would be random rather than equally spaced, and we would probably require more points than are shown in the diagram.) In (a), the relationship (6.10') is used to produce a Breit–Wigner distribution in m; the gradient of the tanget to the curve is such that the values of mass are more densely congregated around M_0 than in regions far from there. The example in (b) uses the relation $\beta = x^2$ to produce a density of x-values which is proportional to x.

relevant distributions is given in Fig. 6.8. Alternatively, the way equation (6.11′) produces distribution (6.7) can be seen simply by writing down a uniform distribution of β-values together with the corresponding x-ones, thus

$$\beta = 0 \quad 0.1 \quad 0.2 \quad 0.3 \quad 0.4 \quad 0.5 \quad 0.6 \quad 0.7 \quad 0.8 \quad 0.9 \quad 1.0$$
$$x = 0 \quad 0.32 \quad 0.45 \quad 0.55 \quad 0.63 \quad 0.71 \quad 0.77 \quad 0.84 \quad 0.89 \quad 0.95 \quad 1.0$$

The distribution in x is more densely populated at the top end, as expected from (6.7).

What are the relative merits of the three different ways of producing a non-uniform distribution? Method (i) suffers from the disadvantage that the final data sample consists of a set of values of the relevant variable, each with its different weight. (Contrast methods (ii) and (iii) which simply provide us with events, each to be treated equally.) This includes the complication that, for the final sample, it is not immediately clear what the equivalent number of complete events is. Furthermore, many of the weights are typically very small and so the corresponding events are effectively rejected, which means that we are wasting computation time.

Method (ii) requires us first to calculate y_{max}. This may not be trivial, especially if y is a function of several variables, rather than of just one as we have assumed. If y_{max} cannot be derived analytically, we estimate it numerically. It is better to over-estimate y_{max} (which simply results in a waste of computer time) rather than to under-estimate it (which produces a bias in the resulting distribution – see Fig. 6.7(b)). Method (ii) also requires two random numbers for each attempt at producing an event. Finally, in many problems, $y(x)$ is small over a large part of the range of x, so that many of the attempts at producing events simply result in that event being rejected.

The best procedure is clearly Method (iii). The only problem is that we have to be able to perform the integration in (6.9); and we must also be able to invert the functional relationship $\beta = \beta(x)$ in order to determine the value of x for a given value of β. If both of these can be done, then this method produces whole events for less computation than the other methods.

These relative merits are summarised in Table 6.3.

Distributions of the type

$$\frac{dn}{dx} = x^{\ell}$$

Table 6.3. *Comparison of the relative merits of the different methods of generating non-uniform distributions*

	Method (i)	Method (ii)	Method (iii)
Do we have to bother about weighted events?	Yes	No	No
How many random numbers needed per attempted event?	1	2	1
Are events thrown away?	Effectively yes	Yes	No
Can method always be used?	Yes	Yes, if we can estimate y_{max}	No. We must integrate, and invert a function

can be produced directly by any of the three methods described above. An amusing alternative (for the case where ℓ is a positive integer) is to generate sets of $(\ell + 1)$ random numbers, and then select the largest of each set; the resulting distribution is x^{ℓ}.

As a final example of a non-uniform distribution, we consider the Gaussian distribution (1.14), which is particularly useful for simulating experimental errors; the difference between observed and true measurements is often approximately Gaussian distributed, hopefully with zero mean. A simple method of generating this distribution is to calculate

$$g = \sum_{i=1}^{n} (x_i - \tfrac{1}{2}), \qquad (6.12)$$

where x_i are random numbers in the range 0–1. Then, by the central limit theorem, g is approximately Gaussian distributed for large values of n. (See Section 1.4.3.) In practice, $n = 12$ is usually large enough. This choice has the additional advantage that not only is the mean zero but the variance is 1 (since the mean and variance of $x - \tfrac{1}{2}$ are zero and $\tfrac{1}{12}$ respectively). This is, of course, not the optimum method of producing a Gaussian distribution, and computer routines exist which incorporate better techniques.

In problems which need pairs of random numbers, two separate calls of the random number generator should produce an uncorrelated pair. In some applications, however, correlated pairs are needed. For example, if a beam of particles is diverging from an upstream approximate focus, the angle θ of a given particle will be correlated with its transverse position y. To produce correlated distributions starting from uncorrelated ones in x_1 and x_2, we can construct linear combinations

$$\theta = x_1 \cos\alpha + x_2 \sin\alpha,$$

$$y = -x_1 \sin\alpha + x_2 \cos\alpha.$$

Then the covariance of θ and y is given by

$$\text{cov}\,(\theta,\, y) = \sin\alpha\cos\alpha\,[\sigma^2(x_2) - \sigma^2(x_1)].$$

Thus correlations will be present provided the errors on x_1 and x_2 are unequal (and both $\cos\alpha$ and $\sin\alpha$ differ from zero). This is readily apparent in considering the distribution of uncorrelated points in the x_1, x_2 plane; if the individual distributions are Gaussian and the errors are unequal, the points will be concentrated approximately within an elliptical region. The transformation to the y and θ variables corresponds to a rotation of axes, and so the ellipse will now be seen with respect to the new variables, i.e. the points exhibit a correlation in y and θ since the ellipse's axes are not parallel to those of the new co-ordinate system.

Our final example in this section is not really a Monte Carlo calculation, although it keeps very much to its spirit. We often plot distributions of quantities which are derived from the variables associated with two (or more) tracks in a given multi-particle event (e.g. the effective mass of the pair, their difference in rapidity or azimuth, etc.). An apparatus with limited acceptance can result in distorted distributions, which can mimic interesting correlation effects that we are trying to discover. In order to estimate the influence of the apparatus without doing a full Monte Carlo simulation taking into account all its details, we can reproduce most of its effect by calculating the distribution of the quantity of interest using pairs of tracks from different events. Such a procedure, however, does not impose the constraints that arise from energy and momentum conservation.

6.5 Typical uses of the Monte Carlo technique

We now outline some more realistic examples of Monte Carlo calculations. With some of the simpler forms of these examples, however, there are other possible ways of solving the problems.

6.5.1 Designing experiments

Before embarking on an experiment, it is necessary to determine whether the suggested experiment can succeed in achieving what it was designed to do. This can be roughly separated into two categories, examples of which are given.

(a) The apparatus

The momentum of a track is usually determined by measuring its curvature in a magnetic field. The curvature in turn is obtained from co-ordinates measured at various positions along the track. Both the finite resolution with which co-ordinates are determined and the multiple scattering produced by the material in the track's path will contribute to the uncertainty in the momentum determination.

The magnitude of these effects can be determined in Monte Carlo fashion. Tracks of chosen momenta are followed through the detection system in the magnetic field. After traversing each thin layer of material, the particle's direction is modified to allow for a random multiple scatter, chosen from a Gaussian distribution whose width depends on the amount of material traversed.† The measured position at each detector is modified to allow for the expected resolution. Then the complete set of measured co-ordinates of a track are used to extract the estimated momentum, and the comparison of these estimates with the known input value then gives the momentum resolution.

If this value, which in general will be a function of the track's momentum, is too large, the design of this part of the apparatus must be changed. For example, the field could be increased, more position measurements made, their accuracy improved or the amount of material in the system reduced.

Although the above description applies to random errors, a similar approach can be used to investigate the influence of systematic errors in the alignment of parts of the detection system or the knowledge of the magnetic field, and hence to determine the maximum allowed tolerance on these.

In a realistic experiment, there could well be several such simulations required to study the response of the various parts of the apparatus.

(b) The experiment as a whole

If an experiment is planned to discriminate between competing theories, it is necessary to find out whether it is capable of so doing. An example of this could be the case considered in Section 4.6.2: if the spin

---

† For a particle of momentum p GeV/c traversing a layer of material of thickness d radiation lengths, the R.M.S. multiple scattering angle θ (in milliradians) is given by $\theta \approx 20\sqrt{d}/p$.

of a decaying resonance is 1, will 30 events be sufficient to rule out the possibility of spin zero?

To test this we generate, say, 100 Monte Carlo 'experiments' each containing 30 events, and distributed according to the theory we are testing, i.e. spin 1, which we are assuming gives a decay distribution proportional to $\cos^2 \theta$. It may also be necessary to include in the simulation any inefficiencies of the apparatus, or its limited experimental resolution.

We parametrise these sets of 'data' in some suitable way, and see how many of the 100 'experiments' give a value consistent with that expected for the isotropic decay distribution of a spin zero resonance. If this number is more than a few, we decide that this particular experiment does not have the power to distinguish between spins 0 and 1, and we try to think of a way of accumulating more than 30 events and/or of improving the angular resolution.

In fact, any proposal to perform a high energy experiment nowadays invariably involves extensive Monte Carlo calculations, in order to convince the selection committee (and the people wanting to perform the experiment) that the project is feasible.

6.5.2 *Testing programs*

We often use large and complicated computer programs to analyse and classify sets of experimental data. In order to test the reliability of such programs, it is useful to use as input some Monte Carlo data of known form, and to check how often our programs succeed in reconstructing the input.

For example, a beam particle interacts in a target, and produces up to six secondary tracks, whose co-ordinates are measured by various planes of counters situated downstream from the interaction point and in a magnetic field. Our analysis programs first have to solve the pattern recognition problem of deciding which co-ordinate pairs from each plane are associated with each other (i.e. correspond to a single particle), and then they must extract from these co-ordinates the track's parameters (i.e. charge, direction and momentum).

We can write a Monte Carlo program to generate secondary tracks, we find whether they intersect the counter planes and if so where (after allowing for multiple scattering in the various materials along the tracks' paths), and introduce a smearing in these measured co-ordinates corresponding to experimental errors, as well as omitting some altogether to allow for counter inefficiencies or for two track hits which are closer than the experimental resolution. Maybe we are even more realistic and introduce a few spurious co-ordinates corresponding to various background tracks passing through our apparatus. These co-ordinates then are used as input to our analysis programs. We are interested in our efficiency for track finding (which may well be a function of the track's momentum and direction as well as how near it is to the next track);† and by comparing the generated track parameters with those actually found, we obtain an idea of the experimental resolution that may be achieved.

Another advantage of such a procedure is that it enables the analysis programs to be written, tested and more or less ready by the time the experimental data are first available.

6.5.3 Contamination estimates

We are also interested in knowing how often complicated analysis programs will tell us that we have an interesting signal when in fact there is only some form of background present. Thus in the example of 6.5.2 our track finding routines not only have the possibility of failing to find real tracks but they may also associate co-ordinates from different tracks or from background sources to produce a spurious track. (In the language of Section 4.6, these are errors of the first or of the second kind respectively.) We can thus investigate this problem using the same Monte Carlo program as is needed for the problem of Section 6.5.2, but here we simply see how frequently the program outputs tracks which we did not generate. As in all problems of this sort which attempt to estimate contamination, the answer will be very sensitive to the exact nature of the input, e.g. the ratio of the background hits to the genuine secondary tracks, the correlations between the secondary tracks' momentum vectors, the experimental resolution, etc.

A similar problem can arise when a detector is used to identify the nature of a particle. Thus muons can be distinguished from other particles by their ability to penetrate large thicknesses of material, while electrons produce characteristic electromagnetic showers. Other particles of known momentum can be identified to some extent by a determination of velocity via measurements of time of flight, Cerenkov radiation, or energy loss. None of these methods provides 100% discrimination between the desired particle type and the background, and it often requires long and complicated

† Thus the track finding efficiency is likely to be a function not only of the parameters of the apparatus but also of the production mechanism of the multi-track events (e.g. phase space, jet production, etc.).

Monte Carlo calculations to determine the efficiency of the proposed method for detecting the wanted particles, and the background of other particles which will be accepted. The latter will be a function not only of the performance characteristics of the detector, but also of the relative fluxes of the wanted particles and the various types of others.

6.5.4 Geometrical correction factors

In an ideal experimental arrangement, all particles produced in an interaction are detected. Real life, of course, is not as simple as that, and it is often a major problem to calculate the detection efficiency of counters that do not completely surround the target where the interactions take place; this detection efficiency is essential in order to convert the *observed* number of events into the physically interesting number of *produced* events. The use of Monte Carlo techniques for solving this problem is elaborated in Section 6.6.1.

6.5.5 Do theory and experiment agree?

Let us assume that we have performed an experiment and that the results do not seem to be in agreement with a particular theory. The question we wish to answer is: 'If the theory is true, how probable is it that we would get a result as weird as ours?' This will then enable us to make a statement of how confident we are that the theory is incorrect.

To do this, we generate, say, 100 Monte Carlo 'experiments' each containing the same number of events as the actual experiment, and distributed according to the theory that we are testing. Once again, inefficiencies and resolution effects can be included in the simulation program. In order to quantify how the experimental distributions agree or disagree with theory, it is possible to parametrise them somehow. Then we simply see whether the value of the parameter for the actual data is typical of that for the collection of Monte Carlo 'experiments', or whether it gives a very different value (see Fig. 6.9). This technique is useful in cases where the errors are non-Gaussian so that the usual statistical theorems and tests cannot be applied.

Of course it is better to test that the whole observed distribution agrees with expectation, rather than to rely on just one parameter. This is probably better done by generating many more Monte Carlo events than real ones, so that in comparing the Monte Carlo distribution with the observed one, the error on the former is negligible.

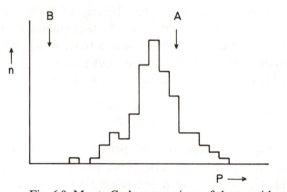

Fig. 6.9. Monte Carlo comparison of theory with experiment. The histogram is for the value of the parameter P as obtained from 100 Monte Carlo 'experiments' generated according to a specific theoretical model. Each of these 'experiments' contains the same number of events as the actual data. If the real experiment produces a value A for the parameter P, then our data are clearly consistent with the theory. On the other hand, a value B indicates that the data are inconsistent with the theory (to the extent that none of the 100 Monte Carlo 'experiments' gives a value for the parameter as small as this).

The evidence for the existence of the gluon is based on such a comparison. The data on hadron production in e^+e^- annihilations at around 30 GeV centre of mass energy and above are inconsistent with predictions from a model based on the mechanism

$$e^+e^- \rightarrow \gamma \rightarrow q\bar{q},$$

followed by the quarks q and $\bar{q}$ then fragmenting into hadrons. Agreement can be achieved, however, by adding a contribution from the process

$$e^+e^- \rightarrow \gamma \rightarrow q\bar{q}g,$$

where all three objects then fragment. These calculations are performed by Monte Carlo methods (see Fig. 6.10).

6.5.6 Resonance or statistical fluctuation?

As a specific example of the situation just considered, we could have plotted a mass distribution in order to search for a new resonance. This would manifest itself as a peak in the distribution at around the relevant mass, whereas in the absence of any such resonances, the distribution would be smoother (see Fig. 6.11). But a peak could be due simply to an

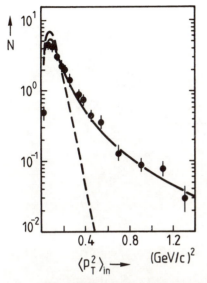

Fig. 6.10. A comparison between data for the process $e^+e^- \to$ hadrons and Monte Carlo calculations for mechanisms via virtual states involving (*a*) a quark–antiquark pair, and (*b*) q$\bar{\text{q}}$ + gluon. The data points are from the TASSO collaboration working at PETRA at centre of mass energies of 27–32 GeV. It shows the distribution in $\langle p_T{}^2 \rangle_{in}$, the mean square transverse momentum for charged hadrons in an event; the transverse momentum is with respect to the event's jet axis, and refers only to that component in the 'plane of the event'. (Further details concerning the definitions of the jet axis and the event plane for multihadron events can be found in the original reference – M. Althoff *et al.*, *Z. für Physik*, **C22** (1984), 307.) The dashed and solid lines are the Monte Carlo predictions corresponding to the intermediate states (*a*) and (*b*) above. The conclusion is that the q$\bar{\text{q}}$ version is insufficient to explain the data, and hence the existence of the gluon is confirmed.

upward fluctuation in the number of events in a couple of adjacent histogram bins from a distribution that was in reality structureless, and while a newly discovered resonant state is a cause for celebration, a statistical fluctuation at best interests no-one and could be the cause of severe embarrassment. So how can we distinguish between them?

We generate 100 Monte Carlo histograms containing the same number of events as the actual experiment, and distributed according to some smooth function approximating the real data (but neglecting the possible resonance). We then add the actual data, shuffle the 101 distributions and show them to our best friend, asking him to pick out the five most likely

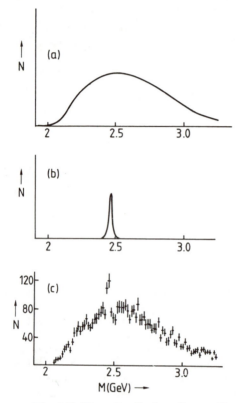

Fig. 6.11. Mass distributions for specific combinations of particles produced in a series of events. (*a*) Without resonance production, the distribution will be some fairly smooth curve. Deviations can arise from statistical fluctuations in the number of events in each bin. Enhancements may also be produced by reflections of resonances in other particle combinations. If particles have been misidentified, an enhancement could correspond to a resonance with different quantum numbers which has been smeared out by the misidentification process. (*b*) Resonance production. The peak is not a delta function because of experimental resolution, and because of the finite width of the resonance. In practice it will sit on top of a background of the type shown in (*a*) because (i) the given resonance may be produced in only a small fraction of the events in the plot; (ii) for each event it may be possible to plot several combinations of the given particle type, while perhaps only one corresponds to the resonance; and (iii) the event sample may be contaminated by background. (*c*) Actual distribution of masses of $\Lambda K^-\pi^+\pi^+$ combinations as produced in interactions of 135 GeV/c Σ^- on beryllium nuclei (S. F. Biagi *et al.*, *Phys. Lett.*, **122B**, (1983), 455). Is the peak at a mass of around 2.46 GeV a statistical fluctuation, or is it a new resonance (possibly a charmed strange baryon A$^+$)?

looking resonances. If he includes the distribution of the actual data, then we are confident at the 95% level that our data do not correspond to a statistical fluctuation and we would be more likely to believe in the resonance interpretation.

In actual practice, we would probably demand a higher confidence level than 95% before being prepared to claim that we have found a new resonance. One reason for this is that, in the course of a typical experiment, we may examine very many (rather than just one) mass distributions, and the chance that one of them may contain a large statistical fluctuation is correspondingly increased.

6.5.7 *Phase space distributions*

We have at last succeeded in obtaining an unbiassed sample of events corresponding to a reaction

$$a+b \rightarrow c+d+e. \tag{6.13}$$

Then, according to the Fermi Golden Rule, the observed distribution dn/dv of the number of events in some kinematic variable (or variables) v may be written as

$$\frac{dn}{dv} = \frac{2\pi}{\hbar} \int |M.E.|^2 \rho \, dw, \tag{6.14}$$

where ρ is the density of states factor and is a function of the kinematic variables, $M.E.$ is the matrix element describing the process, and the integration is performed over the kinematic variables w which are not observed. (See, for example, *Quantum Mechanics* by L. I. Schiff, p. 197 (McGraw Hill, New York 1955).) The density of states factor corresponds to the fact that the final state particles c, d, e, ... want to be produced uniformly in momentum space, in so far as this is consistent with energy and momentum conservation. This can result, of course, in very non-uniform distributions in other variables. Some well-known examples of this are the electron energy distributions in either the free electron gas model of a metal or in the Fermi theory of β-decay. In general that contribution to the experimental distribution arising from the density of states factor ρ is regarded as boring kinematics, while that from the matrix element is interesting physics.

To obtain estimates of the distributions resulting from the factor ρ, we use a Monte Carlo program which generates n-body final states distributed according to the density of states. Several such programs already exist (for example, FAKE and FOWL). We then simply use the generated Monte Carlo events to produce plots of whatever kinematic variable interests us. If this is essentially in agreement with the experimental distribution, there is no reason to invoke any complicated physical theories (via the matrix element) to explain the data.

At high energies, elementary particle reactions show at least one significant deviation from phase space in that the reaction products tend to be produced at small transverse momenta with respect to the beam axis (less than about 1 GeV/c). Many of the observed characteristics of multi-particle high energy reactions follow from this fact. In order to see whether there are any residual effects of interest, the data can be compared with phase space which has been modified to allow for the limited transverse momenta. Programs exist which generate events with tracks produced according to 'longitudinal phase space' in which the density of tracks is uniform in a cylindrically shaped region of phase space which is unlimited in the longitudinal direction (except of course from the constraints imposed by energy and momentum conservation), but cut-off at a suitable value transversely. As with programs that produce events according to ordinary phase space, the user can then construct any plots (e.g. single particle angular distributions, two-particle correlations in rapidity or transverse momentum, mass spectra, etc.) to compare with the actual data.

6.5.8 Matrix elements

In cases where the density of states factor is not sufficient to fit the data, we can modify the prediction by introducing a matrix element which is a function of the kinematic variables. An example of this is elaborated in Section 6.6.2.

6.5.9 Parameter determination

In the example of the previous subsection, we may wish to determine the best value of some arbitrary parameter or parameters which occur in the matrix element. For example, our matrix element may be a linear function of the energy of particle c of reaction (6.13), i.e.

$$M.E. \sim 1 + KE_c,$$

where we wish to determine the value of K.

To do this, we use a Monte Carlo program together with specific values of our parameters in the matrix element to generate calculated distributions (as in Section 6.5.8) which we can compare with the corresponding actual experimental distributions. Then the values of the parameters are varied and the Monte Carlo calculations repeated until the calculated and the experimental distributions are essentially in agreement. Especially when the calculated distributions are approaching the experimental ones, it is desirable to have rather more (say, a factor of 10) Monte Carlo 'events' than actual events, so that fluctuations in the Monte Carlo distributions do not significantly affect the comparison.

This process requires both a large number of Monte Carlo calculations and a minimisation procedure. It is thus liable to take up an enormous amount of computing time, and should be undertaken only when (a) you are really *very* interested in the values of the unknown parameters, and (b) there is no other technique available for calculating them.

6.6. Detailed examples

We now describe in greater detail how the examples indicated in Sections 6.5.4 and 6.5.8 could be implemented in practice.

6.6.1 Geometrical correction factors

We are performing an experiment designed to measure the angular distribution of the reaction

$$a+b \rightarrow c+d, \tag{6.15}$$

using the apparatus shown schematically in Fig. 6.12. The particle c is unstable, and decays in $\sim 10^{-10}$ secs. We assume that spark chambers are used to take photographs of the interactions if and only if both the following conditions are satisfied:

(a) the particle d passes through and is detected by counter D; and
(b) at least one of the decay products of particle c is recorded by counter C.

We want to calculate the geometrical efficiency ε for these conditions to be satisfied, in order that we can correct the observed event numbers to obtain the differential cross-section for reaction (6.15). Since ε in general varies with production angle, not only the absolute value of the cross-section but also the shape of the angular distribution is sensitive to our calculation

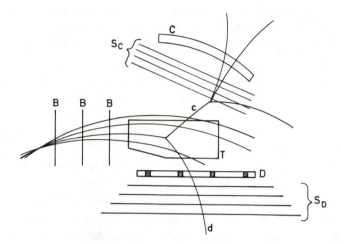

Fig. 6.12. A schematic diagram representing the apparatus used for studying the reaction $a+b \rightarrow c+d$. A magnetic field is used to measure the charged particles' momenta from their curvatures. The beam particles enter the apparatus from the left, and are diverging from an upstream focus. One of them interacts in the target T. If particle d is detected in counter D and a decay product of c in the counter C, the spark chambers B, S_D and S_C are triggered to take photographs in order to enable the particle trajectories to be reconstructed. The shaded areas of D correspond to dead regions in which the counter is insensitive. The detection efficiency ε of the apparatus as a function of scattering angle (but integrated over all the other variables) is studied by a Monte Carlo method.

of ε. We are also interested in our experimental angular resolution; if this is poor, we may be unable to see fine details of the angular distribution (e.g. what happens very close to the forward direction, possible dips, etc.).

There may be many factors complicating the problem and making an analytic solution impossible.

Complication (*i*)
If the target were infinitesimal in size, all interactions would occur at the same space co-ordinate, which would simplify the calculations. But real targets are finite in extent, and the efficiency varies as a function of position in the target.

Complication (ii)

If all beam tracks followed the same trajectory, life would still be relatively simple in that there would only be one variable characterising the interaction point (the co-ordinate along the beam direction). But actual beams have a finite spatial extension across the face of the target. Combined with a funny target shape, it may also mean that different beam particles pass through different thicknesses of target. Furthermore, the beam's lateral positions may be correlated with its direction at the target position; for example, the beam may be diverging from a focus somewhere upstream.

Complication (iii)

The beam and charged secondary particles curve in the magnetic field and slow down as they pass through the material of the target and detectors.

Complication (iv)

The trigger counters C and D can have complicated shapes and include dead regions, which are insensitive to particles. The solid angles that they subtend at the target vary with the position of the interaction vertex within the target.

Complication (v)

The efficiency of counter D for recording particle d may vary with the energy of d (and hence also with the scattering angle).

Complication (vi)

The counter C may fail to record anything either because the particle c did not decay until it had passed through the counter C, or because although it decayed upstream of C, its decay products missed the counter. Both these factors depend on the momentum of particle c, and hence again on the production angle.

Complication (vii)

The particles c or d or the decay products of c may interact before they reach the relevant counter, and hence fail to produce a trigger signal.

We solve this problem by integrating over all the uninteresting variables in Monte Carlo fashion, in order to obtain the efficiency as a function of production angle. The basic steps are:

Step (i)

We set up the experimental distribution of beam tracks in space and in angle, using one random number for each co-ordinate. If these variables are correlated, we first transform to new ones which are uncorrelated (cf. Sections 3.4(b) and 6.4). We generate the independent distributions in the new variables, and then transform back to get the correlated space–angle distribution of the beam.

Step (ii)

We must now determine whether a given beam track interacts in the target or simply comes out the far end. In a crude approximation, we neglect multiple Coulomb scattering and consider that the interaction cross-section of the beam with the target nuclei is independent of the position in the target (either because the energy loss across the target is small, or because the cross-section is essentially energy independent). Then the interaction position x is distributed exponentially

$$\frac{\mathrm{d}p}{\mathrm{d}x} = \frac{1}{\lambda}\,\mathrm{e}^{-x/\lambda}, \tag{6.16}$$

where p is the probability of the interaction being in a given range of x, and λ is the mean free path. We can generate interactions distributed according to (6.16) by Method (iii) of Section 6.4. If the generated value of x is less than the target thickness, we have an interaction, and we proceed. Otherwise the beam has passed through the target and we return to Step (i) for the next beam track.

More realistically, and at the expense of much more calculation, we have to allow for the beam track curving in the magnetic field, slowing down and having an energy dependent cross-section (and hence mean free path), undergoing multiple Coulomb scattering, possibly leaving the sides of the target, etc. Thus we divide the path into small segments, in each of which the beam has a small probability of interacting; otherwise it multiple scatters, is deflected by the field and loses energy in passing through the material of the target. The last two are calculated analytically or numerically; the first two are calculated by a Monte Carlo method. The process is continued until either the beam track leaves the target or interacts.

Step (iii)

Given that the beam track has interacted, we must choose the centre of mass scattering angle θ. We can select it randomly such that $\cos\theta$ is uniformly distributed. Alternatively, since we are interested in obtaining a more accurate estimate of the efficiency ε in that region of θ where the experimental data are congregated, we could choose the $\cos \theta$ distribution of the Monte Carlo events to resemble that of the actual data. Exact agreement is not necessary in order to calculate ε as a function of θ. If, however, we wanted to calculate the value of ε averaged over all production angles from the Monte Carlo data alone,† then the exact form of the Monte Carlo $\cos \theta$ distribution is important.

It is also necessary to choose the azimuth of the scattering plane (i.e. the orientation about the beam direction of the plane of the final state particles) at random; although physics is independent of this,‡ geometrical efficiency factors are not.

Step (iv)

We must check that particles a and c (or the latter's decay products – see Step (v)) emerge from the target successfully without interacting. This is essentially the same procedure as step (ii) for the beam track, except that here, if we have an interaction, we must go back to Step (i) to find a new beam track.

Step (v)

We must let particle c decay randomly in time according to the usual exponential distribution. If this decay occurs beyond the counter C, this interaction does not trigger the system,§ and we return to Step (i). If the decay occurs upstream of counter C, we select the momenta at random according to 3-body phase space, and then choose the decay angles of c at random (and uniformly, assuming particle c is unpolarised or unaligned), to ensure that at least one of the decay particles passes through the counter. Otherwise return to Step (i).

Step (vi)

Next we must check that particle d passes through counter D, and if so, whether the counter records the particle. Thus for the actual momentum of d, the detection probability may be 79%. Then we generate a random number r_d and check whether

$$0.79 > r_d. \tag{6.17}$$

---

† In general, however, it is possible to obtain ε as $(1/n)\,\Sigma\varepsilon(\theta_i)$ where θ_i is the observed value of the scattering angle, $\varepsilon(\theta)$ is the Monte Carlo calculated efficiency, and the summation extends over the n events of the actual data.

‡ We assume the beam and target particles are unpolarised.

§ In some circumstances, particle c can decay beyond counter C, but one of its decay products can still pass backwards through the counter. If this sort of configuration is possible, then the Monte Carlo program should allow for it.

If this is so, we decide that the particle d has been recorded, and we have a valid trigger. Otherwise we must return to Step (i) to generate another beam track.

Step (vii)

We then find where the beam track, particle d and the decay products of c had passed through the spark chambers B, S_D and S_C. To obtain the experimentally recorded co-ordinates, these intersection points are smeared by the experimental resolution σ on the measured co-ordinates; for this we add to each calculated intersection point a random amount obtained from a Gaussian distribution of mean zero and variance σ^2 (generated, for example, as described at the end of Section 6.4). We assume that the spark chambers subtend larger solid angles than their respective trigger counters, or there will have to be a further step in the geometrical efficiency calculation.

Step (viii)

At this stage we have one successful event triggering the system, out of the several events generated at Step (iii), whereas we want a sensible number in each angular range of θ. Ideally, the number of successful trigger events from the Monte Carlo program should be an order of magnitude larger than the number of real events, in order that the statistical error on the geometrical efficiency calculation is smaller than that arising from the number of observed events. So we return to Step (i) and repeat the whole process as many times as is considered desirable or until we have run out of computer time, whichever is sooner. Then we calculate the geometrical efficiency $\varepsilon(\theta)$ of our experimental configuration at the given beam momentum as

$$\varepsilon(\theta) = \frac{\text{No. of events out of Step (vi)}}{\text{No. generated in Step (iii)}}, \tag{6.18}$$

where both numbers refer to a specific angular range of θ. This may have the form of Fig. 6.13. It is this efficiency which is used to convert the observed number of events into the number of interactions actually produced in the target.

We stress that this is simply a calculation of the *geometrical* efficiency. There may be several processes involved through which events have to pass before they appear in a final experimental distribution, and each of these has its own inefficiency for which we must correct. These include, for example, the spark chambers themselves; and the data processing and

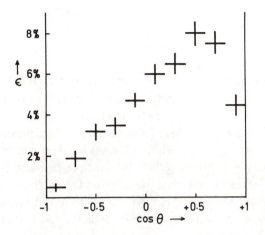

Fig. 6.13. The geometrical efficiency ε of the trigger system for the apparatus of Fig. 6.12. This is calculated by a Monte Carlo program as the ratio of events which succeed in triggering the system to the number of interactions produced, for that particular range of centre of mass scattering angle θ. This efficiency is then used to obtain the number of events *produced* in the actual experiment (as a function of θ) from the *observed* number of events.

analysis programs of the type described in Section 6.5.2, which reconstruct the tracks in space from the measured spark chamber co-ordinates and then perform a kinematic fit to the measured spark chamber co-ordinates.

Finally we can use our Monte Carlo data to obtain the angular resolution of our apparatus. This is obtained simply from the variance of the distribution of

$$\Delta = \theta_c - \theta_g,$$

where θ_g is the generated angle and θ_c is the angle calculated for the same Monte Carlo event from the fitting programs. The resolution could well be a function of θ. The main contributions to the angular resolution come from the measurement errors on the tracks as detected in the spark chambers, and from the multiple Coulomb scattering (for low momentum tracks). The kinematic fitting procedure should result in improving the resolution, unless there are any features in the program which distort the data.

6.6.2 Physics

In order to remember that it is physics in which we are really interested, we consider the problem outlined in Section 6.5.8 of obtaining the prediction of a specific model to compare with our experimental data on the reaction

$$a + b \rightarrow c + d + e. \tag{6.19}$$

As seen from eqn (6.14), the predicted experimental distribution is given in terms of the matrix element and the density of states factor; the Monte Carlo technique here once again helps us to do the integration.

We use a phase space program (like FAKE) to generate events uniformly according to the density of states factor ρ. The specific physics theory enters in the form of the matrix element in (6.14). In principle this can be incorporated by using either of the first two methods described in Section 6.4; below we adopt Method (i) (i.e. we are going to use events with their appropriate weights, with the output from the calculation consisting of weighted histograms).

The model we wish to check predicts that for reaction (6.19), the matrix element is given by

$$|M.E.|^2 = g^2 \, e^{-\lambda p_T^2}, \tag{6.20}$$

where g is a coupling constant whose value will not affect the shape of any experimental distribution, p_T is the transverse momentum of particle c, and λ is a positive constant. Someone has produced two histograms of the real events of reaction (6.19). Both are of an angle ϕ defined by

$$\cos \phi = \frac{(\mathbf{a} \wedge \mathbf{c}) \cdot (\mathbf{b} \wedge \mathbf{e})}{|\mathbf{a} \wedge \mathbf{c}||\mathbf{b} \wedge \mathbf{e}|}, \tag{6.21}$$

where $\mathbf{a}$, $\mathbf{b}$, $\mathbf{c}$ and $\mathbf{e}$ are the momentum vectors of the respective particles as seen in the rest system of particle d. The first histogram is for all events of reaction (6.19), while the second is only for those events for which the energy E_c of particle c and the angle θ_d of particle d are larger than some chosen values. The two histograms look somewhat surprising, and we want to know what is the theoretical prediction for them.

In order for a standard program like FAKE to produce the correct sort of events, we must specify the beam particle and energy, the number of final state particles and their masses, and how many events are to be generated.† The program also requires further information concerning the

---

† As usual, this should exceed the number of real events by an order of magnitude (unless it is already clear with fewer events that the theory is very different from the experimental distribution). The relevant number here is the effective number of events after weighting, which of course is less than the number of events which we request to be generated by the program.

histograms which are to be produced at the end of the generation of the specified number of events.

FAKE contains a user subroutine in which the events are presented to us one at a time. Each event consists of 4-vectors corresponding to the beam particle and the outgoing secondaries (in this case c, d, and e). It is now up to us to calculate from these vectors the kinematic quantities of interest (ϕ, p_T, E_c and θ_d); for this there usually are useful subroutines which perform Lorentz transformations, the standard vector manipulations, etc. For each event we calculate a weight w, which is the value of $|M.E.|^2$ as specified in eqn (6.20). Then we enter our quantity ϕ of eqn (6.21) with its weight w in the correct bin of the first histogram for each event, but in the second histogram only if this event satisfies the selection criteria on E_c and θ_d; this check we must perform in the user subroutine. As well as entering the events into histograms as appropriate, it is highly desirable to print out in detail the first few events (i.e. the 4-vectors of the particles and the kinematic quantities we need) as an aid to debugging and checking out the program; this would be difficult if all that is available is the final histograms.

At the end of the run, the histograms requested are printed and can be compared with the experimental data. If the Monte Carlo distributions do agree with the corresponding experimental ones, how excited should we get? It might, of course, have been expedient to produce the corresponding plots with the matrix element set equal to unity. If the agreement is still good, our excitement should be somewhat tempered by the knowledge that even a simple phase space factor produced as good agreement with the data as does our theory. Also there is the fact that there is an infinite number of experimental distributions that can be produced from the kinematic quantities of reaction (6.19), and if the theory is correct, the predictions for all of them should be in agreement with the data; so far we have checked only two.

The reader need not be too disturbed by these philosophical points, however, since the more likely situation is that the Monte Carlo and experimental histograms disagree with each other, so we have the other problem of thinking of a new model which will provide a better description of our data.

6.7 Non-physics applications

We end this chapter on Monte Carlo techniques with two examples from outside the field of nuclear physics.

6.7.1 *How to save yourself a fortune*

Here the Monte Carlo technique is applied almost self referentially to the Monte Carlo problem. Of those who are attracted by gambling, almost everyone has devised a 'system' by which they can win a vast amount of money. It would seem prudent first to test the performance of this system in an environment where no money is required.

Let us assume that the game involved is roulette. Then we arrange our random number generator to produce for us a series of results of spinning the roulette wheel; the numbers 1–36 and an extra blank are produced each with equal probability of 1/37.† We then apply our system to a series of spins. The system must incorporate decisions on what number or groups of numbers to bet on, how much to wager at each throw, and when to stop. It is also important to decide how much money we have available at the beginning of the session.

We can then see what is the outcome of a series of games. This can be a profit or loss at the end of the sequence, or our losing all our money at some intermediate stage; we ignore the possibility of our winning so much that the bank runs out of money. By repeating this sequence a large number of times, for which a computer is highly desirable, we can determine the probability distribution of the financial outcome. (It is the distribution, rather than simply the mean, which is of significance in judging how good a system is.) By repeating the whole procedure for other systems, we may eventually decide which we favour, or whether it may not after all be better to avoid the Monte Carlo casino.

6.7.2 *The Garden of Eden problem*

Let us assume that we are trying to populate the Garden of Eden with human beings. We are provided with Adam and Eve as original inhabitants. The population is increased by childbirth, but decreased by death. We are interested in the likely population after 200 years, and would be very unhappy if at that stage there was no one left alive (or if all survivors were

† We are assuming that the roulette wheel is being operated in a fair, unbiassed manner.

of the same sex). Thus, if we are allowed to choose the value of the ratio of male-to-female children born as a free parameter, what is its optimum value in order that there is the highest chance of a surviving population?

In order to specify the problem more precisely, it is necessary to add a few more assumptions.

(i) The problem commences in year zero with Adam aged 15 and Eve 12.

(ii) Women aged 12–40 are potentially capable of having children.

(iii) Women on reaching the age of 12 get married to any unmarried man of 15 or over; if an eligible man appears only later, then they get married at that stage.

(iv) There are two categories of women. The faithful type have a probability of 0.25 per year for having children if they are married. The unfaithful variety have a yearly probability of $0.75(1-0.9^n)$, where n is the total number of men aged 15 and over, plus a further 0.25 if they are married. 30% of women are faithful, and 70% unfaithful. Women are absolved from having children for two years after any delivery.

(v) At each pregnancy, the probability of m children being born is proportional to $e^{-4(m-1)}$ (for $m \geq 1$). The probability of any child being male is p, the parameter we are going to vary.

(vi) The probability of surviving childbirth is 95% for the mother, and similarly for each child.

(vii) The probability of surviving any year (apart from the effect of childbirth) is 99% for women, but a lower 97% for men who lead a more dangerous existence.

The idea of finding an analytic solution to the above problem is horrendous. For a question as important as the survival of the human race, a solution should be found, and so Monte Carlo techniques must be used. In outline, we start off with Adam and Eve, and consider one-year-steps. We use random numbers to decide whether Eve should be categorised as faithful, whether they have any children (and if so of what sex) and whether they die. In subsequent years, we arrange such marriages as are possible. It is also necessary to increment everyone's age by one year. After we have repeated this process 200 times, we see whether we have a population of at least one male and one female, which would count as a success. We then repeat this procedure a large number of times, and obtain the probability P of a surviving population for the value of our male/female parameter p. It now merely remains to go over the whole calculation for a series of values of p in order to choose the one that results in the largest value of P.

We leave the detailed implementation of this program as an exercise for the reader.

Problems

6.1　Check that your computer (or calculator) can generate 10 000 (or 100) numbers that satisfy some basic criteria for random numbers.

(a)　The first number is not repeated in the sequence.

(b)　Plot a histogram to ensure that the distribution of random numbers in uniform over their range.

(c)　In order to look for possible correlations between consecutive random numbers, plot a histogram to find the distribution of the 5000 (or 50) differences between pairs of adjacent random numbers in the sequence. Does this have the expected shape?

6.2　Find a value of π by a Monte Carlo simulation of the following situation (Buffon's needle). A piece of paper is ruled with a series of lines parallel to the x-axis and with unit spacing. A needle of unit length is spun and allowed to fall on the paper so that its centre is randomly and uniformly distributed over the range -1 to $+1$ in y, and its orientation is also random and uniform in angle. The probability that the needle lies on a line is $2/\pi$.

Calculate how many throws of the needle are required in order to evaluate π to an accuracy of ± 0.1. Simulate this number of throws of the needle, find what fraction lie on a line, and hence determine π. Is it suitably close to the correct value?

6.3　Produce some Breit–Wigner distributions with $M_0 = 784$ MeV and $\Gamma = 12$ MeV by Monte Carlo methods as follows.

(i)　Generate 1000 events uniformly in mass over the range 748–820 MeV, and then weight each event by the factor

$$B = \frac{\Gamma/2}{(m-M_0)^2+(\Gamma/2)^2}.$$

(ii) Generate events as in (i), but instead of weighting them keep them in the distribution only if

$B/B_0 >$ another random number.

where B_0 is the value of B when $m = M_0$

(iii) Generate 1000 events uniformly in the angle β over the range $-90°$ to $+90°$, and then transform to mass m by

$$\tan \beta = \frac{m - M_0}{\Gamma/2}.$$

Use a maximum likelihood method to obtain a best estimate of M_0 and Γ using as input data the mass values generated by your Monte Carlo program in Method (iii).

6.4 Investigate the extent to which small samples of a random variable do or do not provide good estimates of the population variance as follows:

Generate 100 samples, each of four measurements of a random variable, which is assumed to be Gaussian distributed with mean zero and unit variance. Obtain an estimate s^2 of the variance for each sample, using formula (1.3″) with the constant $k = -1$. Plot the distribution of the 100 s^2 estimates, and calculate their mean m and also the value of v^2, the mean square deviation of the s^2-values from unity. Repeat the calculations for $k = 0$, and for $k = +1$.

Are the values of m and of v^2 as expected? (See Problem 5.12.)

Background subtraction procedures

Let us assume that we want to determine some parameters characterising a resonant state as produced in a particular reaction. This could, for example, be the fraction of its decay particles that are kaons, a constant in the decay distribution of the resonance, the polarisation of one of its decay products, etc. We are frequently faced with the situation that the resonant peak as experimentally observed in the relevant mass spectrum sits on a non-resonant background. Thus Fig. A 1.1 could represent the variation with energy of the cross-section for the reaction

$$a+b \rightarrow C^* \rightarrow d+e, \qquad (A\ 1.1)$$

where C^* is the intermediate decaying resonant state; or it could be the effective mass spectrum of f and g in a reaction

$$a+b \rightarrow f+g+h, \qquad (A\ 1.2)$$

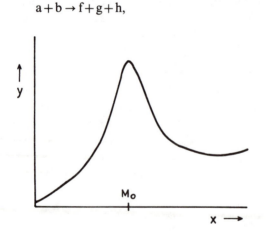

Fig. A 1.1. A spectrum indicating a resonance of mass M_0. For reaction (A 1.1) this would be a plot of the reaction cross-section y as a function of the centre of mass energy x, and the resonance C^* is coupled to both the $a+b$ and the $d+e$ systems. For reaction (A 1.2), we plot the number of events y as a function of the effective mass of the $f+g$ system into which the resonance J decays.

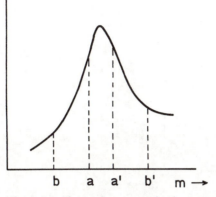

Fig. A 1.2. If we select events in the narrower mass region *aa'*
extending between $M \pm \Gamma/4$, the resonance to background ratio is
reasonably good (4.6:1 in this particular case), but we include only
30% of the events of the resonance. For the larger mass region *bb'*
($m \pm \Gamma$), the signal to background ratio is worse (2.8:1) but we accept
70% of the resonance's events.

in which case the bump corresponds to the two-stage process

$$a + b \rightarrow J + h, \quad J \rightarrow f + g, \tag{A 1.2'}$$

and the smooth background could be due to the direct production of
reaction (A 1.2). Another possible source of background in either histogram
is simply from interactions of a different type which our detectors fail to
discriminate from the reaction that we thought we were studying. In either
case if we simply select for further study events in the region corresponding
to the peak of the resonant plot, we accept background events as well, and
hence any parameter which we determine using these events is characteristic
of the 'resonance plus background'. Then the problem is how to correct
for the effect of the background, so as to determine the properties of the
resonance alone.

Method (i)

If the background within a chosen mass region around the resonance is
small, we may try to pretend it is not there at all, and just forget about
it. We must of course compromise on how we define the mass region within
which we accept events: the narrower this is, the better our signal/back-
ground ratio, but the smaller the number of accepted events and hence the
worse is the statistical accuracy (see Fig. A 1.2).

Clearly it is not worth determining the parameters of the resonance to

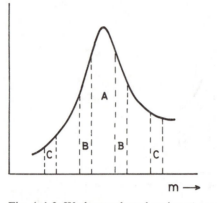

Fig. A 1.3. We have selected region *A* as our resonance region. We choose control regions (for subtracting the effects of background) by selecting two regions, one on either side of the peak, and each half as wide as *A*. With the control regions *B*, we subtract a large number of resonant events, thereby decreasing the statistical accuracy of the answer; in control regions *C*, the resonance contribution is smaller, but the approximation of a linear variation of the background becomes worse.

very high statistical accuracy in such a method. Thus the better the statistical accuracy of the data, the lower should be the upper limit on the background that we are prepared to tolerate if we are to apply this method.

Method (ii)

If we may assume that the background varies linearly across the mass region of interest, and that its interference with the resonance is neglible,[†] then we can use control regions on either side of the resonance region to estimate the effect of the background (see Fig. A 1.3).

If we plot some distribution *d* for the resonance region, we assume that it is given by

$$d = \frac{n_R d_R + n_B d_B}{n_R + n_B},\tag{A 1.3}$$

where n_R and n_B are the numbers of resonant and background events[‡] in the so-called resonance region, and d_R and d_B are the distributions that we would obtain for pure resonance and for pure background respectively.

† This is certainly the case if the background consists of a different misidentified reaction. If the background really is from the same final state, then whether or not there is significant intereference depends on the physics of the processes in question.
‡ When interference effects are important we cannot divide up the total number of events in this simple way.

Because of our assumption concerning the form of the background, the distribution for the control regions is

$$d' = \frac{m_R d_R + n_B d_B}{m_R + n_B},$$ (A 1.3′)

where the number of resonant events m_R is smaller than before, but the number of background events n_B is the same. Thus when we subtract the distribution of the control regions from that of the resonance region, we obtain

$$d_s = \frac{(n_R + n_B) d - (m_R + n_B) d'}{n_R - m_R}$$

$$= d_R,$$ (A 1.4)

i.e. this distribution should, provided our assumptions are justified, be that of the pure resonance. Note that we derived this result without having to assume that the distributions d_R and d_B have similar functional forms.

We can then use any of the standard methods to determine our unknown parameters from this background-subtracted distribution. In applying the method of moments, we simply use events in the resonance region with weight $+1$ and those in the control regions with weight -1. This is equivalent to determining separately the moments M and M' from the n and n' events in the resonance and control regions respectively, and then determining the resonance's moment M_R in terms of these as

$$M_R = \frac{nM - n'M'}{n - n'}.$$ (A 1.5)

It may, however, be incorrect to use a formula like (A 1.5) for actual parameters which are given in terms of the moments since (a) the relation between the moment and the parameter is not necessarily linear, and (b) the distributions d_R and d_B may not be of the same form, and hence the parameter in question may not be relevant for the background events.

As already pointed out in Comment (viii) of Section 4.4.3, there can be difficulties in applying likelihood methods to this background-subtracted technique.

It is important to remember that in the background-subtracted distribution, the statistical error on the number of events h in a given histogram bin is not $\sqrt{h}$, but

$$\left.\begin{aligned} \sigma_h &= \sqrt{(\sigma_r^2 + \sigma_b^2)} \\ &= \sqrt{(r + b)} \\ &> \sqrt{h}, \end{aligned}\right\}$$ (A 1.6)

where r and b are the number of events in the corresponding bin for the resonance and background regions respectively (i.e. $h = r - b$). Indeed h can be negative, while the error is still respectable.

When selecting the sizes of the mass regions in this method, we are faced not only with the difficulty described in Method (i), but also with the fact that the control regions should be close enough so that the linear background approximation is still reasonable, and yet not so close that they contain almost as many resonant events as does the central region. One possibility is to have the control regions slightly separated from the central region (see Fig. A 1.3).

Because of these problems associated with the choice of the sizes of the regions, we are encouraged to look for other methods which avoid specific choices and which attempt to use efficiently all the events in the resonant peak.

Method (iii)

This is an attempt to use the method of moments for determining the resonance parameters. Consider first dividing the mass range across the resonance into n bins, and determining the moment M_i of the same function in each of them. We assume that these moments should behave as

$$M_i = f_i M_R + (1 - f_i) M_B, \tag{A 1.7}$$

where M_R and M_B, the moments for the pure resonance and pure background respectively, are assumed independent of mass; and f_i is the fraction of resonance in the ith histogram bin, and is assumed known as a function of mass from the fit to the mass spectrum of Fig. A 1.1. Thus if the fitted mass spectrum is $BW(m) + L(m)$†, then

$$f_i = \frac{BW(m_i)}{BW(m_i) + L(m_i)}. \tag{A 1.8}$$

Then we wish to determine the two unknowns M_R and M_B from the n experimentally determined moments, one from each mass bin. To do this, we minimise

$$S = \Sigma \left(\frac{M_i - f_i M_R - (1 - f_i) M_B}{\delta M_i} \right)^2. \tag{A 1.9}$$

† E.g. a Breit–Wigner resonance, and a linear background.

This has as solution

$$M_R = \frac{Rb - Bc}{rb - c^2}$$

and (A 1.10)

$$M_B = \frac{Br - Rc}{rb - c^2}$$

where

$$R = \sum \frac{M_i f_i}{\delta M_i^2},$$

$$B = \sum \frac{M_i(1 - f_i)}{\delta M_i^2}$$

$$r = \sum \frac{f_i^2}{\delta M_i^2}$$ (A 1.11)

$$b = \sum \frac{(1 - f_i)^2}{\delta M_i^2}$$

and

$$c = \sum \frac{f_i(1 - f_i)}{\delta M_i^2},$$

The inverse error matrix is determined from the second derivatives of S as

$$\begin{pmatrix} \delta M_R^2 & \mathrm{cov}(M_R, M_B) \\ \mathrm{cov}(M_R, M_B) & \delta M_B^2 \end{pmatrix}^{-1} = \begin{pmatrix} r & c \\ c & b \end{pmatrix}.$$ (A 1.12)

Now we wish to go to smaller and smaller bins, until no bin contains more than one event. Then we speculate that we retain the formulae (A 1.10) for M_R and M_B, except that the summations are now over the individual events, with M_i being the contribution of an individual event to the moment. The δM_i are all set equal to each other for the time being. We determine their absolute value (which is needed for the evaluation of the error matrix, but not for the values of M_R and M_B) by equating S to its expectation value of $m - 2$, when m is the number of events.

As we now have estimates of M_R and M_B, we could try to discover how δM is expected to vary with mass, and then use this to replace our previously assumed constant value, i.e. we iterate to get a slightly improved solution for M_R and M_B. But we consider this to be contrary to the spirit of simplicity of moment calculations, and expect to use just the first estimate.

The values of M_R and its error are then used to calculate the parameter in which we are interested.

Method (iv)

Another method which does not require mass bins involves writing the distribution of interest as

$$D(\theta, m) = d_R(\theta)\, n_R(m) + d_B(\theta)\, n_B(m), \tag{A 1.13}$$

where $d_R(\theta)$ and $d_B(\theta)$ are the distributions in the variable(s) θ for the resonance and background respectively, and $n_R(m)$ and $n_B(m)$ are the number of resonant and background events as a function of mass. Then using events over a range of mass values, we use a likelihood method† to determine simultaneously the parameters of the resonance and background distributions d (and perhaps of $n(m)$ too).

The two functions d_R and d_B need not be of the same form, but at least they must be specified to within some unknown parameters. The mass functions n_R and n_B would typically be a Breit–Wigner and a polynomial respectively, although this method is much more general and would allow almost any functional form in principle. This method can also be extended rather simply to allow for interference effects between background and resonance amplitudes.

The main disadvantages of this method are that (a) often a suitable functional form for describing the background is not known *a priori*, and may be difficult to determine experimentally; and (b) eqn (A 1.13) may involve a large number of parameters, and hence the maximisation of the likelihood function may be difficult.

The moral of this appendix should by now be clear – it is much easier to study a resonance when the background is small.

† See also p. 93.

APPENDIX 2

The number of constraints in kinematic fitting

In this appendix, we return to one particular aspect of the kinematic fitting problem of Section 4.6.2(b), namely how many constraints there are in various situations. (See also Sections 5.2.2 and 5.2.3.)

Already in Section 1.3, we saw that constraints are useful as checks, and the more constraints there are, the better. As a very crude example, we consider the case of several quantities, which are uncorrelated and are restricted to lie in the range -1 to $+1$, and which according to the constraints applying to the particular problem should all be zero. If we measure each quantity to an accuracy of ± 0.25, the probability of one constraint being satisfied, even when the relevant hypothesis is incorrect (i.e. an error of the second kind) is about 25%, but the probability of four such constraints being simultaneously satisfied when the hypothesis is wrong is only about 0.4%.

The examples we give below are all taken from high energy elementary particle physics. We assume that the experiment employs a visual detector such as a bubble chamber, in which the charged particles are detected and in general will have their momentum vectors measured;† the production vertex and (except in Example (xiii)) any decay vertices are also assumed to be observed. The principles, however, apply equally to other situations which give rise to track vectors of similar configurations.

Example (*i*)
Consider a reaction

$$pp \rightarrow pp\pi^+\pi^-, \tag{A 2.1}$$

in which the momentum vectors of all the particles (i.e. the four final state particles and the beam – the target is at rest) are measured, and there are

---

† The energies of neutral particles are sometimes determined via their energy deposition in a calorimeter. Then, for an assumed particle type and production position for the neutral particle, its momentum vector is obtained (in general with poorer precision than for a typical charged particle).

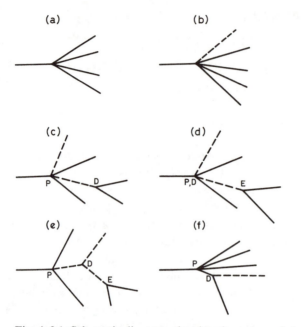

Fig. A 2.1. Schematic diagrams showing the measured charged tracks (solid lines) and the undetected neutral ones (dashed). No attempt has been made to show the curvature of the charged tracks in the magnetic field. The beam track is in each case incident from the left. (*a*) Reaction (A 2.1) with four outgoing charged tracks, none of which decays. (*b*) Reaction (A 2.2) with four charged and one neutral track. Again none of the outgoing particles is seen to decay. (*c*) The reaction sequence (A 2.8), in which a neutral particle is produced at *P* and decays at *D* into two charged tracks. Also produced at *P* are two charged and one other neutral particle. (*d*) Reaction (A 2.12) involving a Σ°, whose lifetime is so short that it appears to decay at the same point that it is produced (points *D* and *P* respectively). Its decay products are a γ (the unseen track from *D*) and a Λ°, which is seen to decay into two charged tracks at *E*. (*e*) Reaction (A 2.15) involving a Ξ°, which is produced at *P*, and which decays at *D* into a π° (unseen) and a Λ°, which subsequently is seen to decay at *E*. (*f*) Reaction (A 2.18), in which a Σ^- produced at *P* decays at *D* into a charged and a neutral track (π^- and neutron respectively). Three other charged particles are produced at *P*.

no complications from observed decays. (See Fig. A 2.1(*a*).) Then there are four constraints – energy and the three components of momentum must be conserved in reaction (A 2.1). If these are all satisfied to within one or two errors (which is equivalent to obtaining a satisfactory low value of S_{min} in eqn (5.43)), then an error of the second kind is probably very unlikely (except for the circumstances of Examples (v) or (vi)).

Example (ii)

If in Example (i), m of the track variables are unknown then the number of constraints is reduced by m. Thus if, for example, a momentum is undetermined (because the track is travelling parallel to the magnetic field, or because it scatters after such a short distance that its curvature cannot be determined), then one constraint is lost.

Example (iii)

In a reaction such as

$$pp \rightarrow p\pi^+\pi^+\pi^-n, \tag{A 2.2}$$

one of the particles (in this case the neutron) is completely unseen (because it is neutral and does not interact or decay visibly – see Fig. A 2.1 (b)). Then three of the conservation equations are used up in determining the unknown 3-momentum vector of the unseen particle, and so only one constraint is left as a check. Thus our χ^2 check on S_{min} is equivalent to asking: 'Is the missing mass† in our reaction equal within errors to that expected for the hypothesised neutral particle, i.e. to that of the neutron?' Especially when the experimental measurement errors are large, there is a danger of making an error of the second kind. Thus in order to obtain a relatively pure sample of such events, it is sensible to use a higher rejection probability cut (say 10%) for reactions involving only one constraint than for a 4-constraint reaction (where a 1% cut may be more appropriate). Even so, when testing several possible reactions each involving a missing neutral, we may find that an event gives an acceptable probability for more than one reaction, and hence the event is ambiguous.

If a 4-constraint and a 1-constraint hypothesis both result in probabilities of greater than 10%, the 4-constraint reaction is in general intrinsically more probable, in that it is more difficult to satisfy several constraints simultaneously if the hypothesis is incorrect.

---

† From the measured quantities, we can calculate the imbalance of momentum and of energy, which are carried away by the neutral particle. From these, we can calculate the apparent mass (called the missing mass) of the undetected system. The missing energy must of course be positive, or such a calculation is not sensible.

Example (iv)

If in the case of Example (iii), the magnitude of the momentum of one of the charged particle is unmeasured, then there are four unknown quantities in all, which require all the four conservation laws for their determination. Thus there are no constraints left to check whether the observed event is consistent with the hypothesised reaction. Sometimes there will be no real solutions to the four equations for the four unknowns (for example, the observed track energies may be too large to permit an extra unseen neutral whose energy is necessarily positive) and hence the hypothesis can be rejected. But in general a solution exists (or, to be more precise, two solutions since one of the equations is quadratic) and no check of the equality type is possible. In such circumstances, the event is likely to be ambiguous, with several possible reactions each involving one unseen particle. It may thus be sensible to reject such events, and to weight the remaining events in order to correct for this loss.

Example (v)

If we have observed an interaction with four outgoing charged tracks, we may wish to distinguish between the hypotheses

$$\text{pp} \rightarrow \text{pp}\pi^+\pi^- \tag{A 2.3}$$

and

$$\text{pp} \rightarrow \text{pp}\text{K}^+\text{K}^-. \tag{A 2.3'}$$

Each of these reactions involves four constraints, and so it sounds as if the separation should be very easy. But in fact, it is not so. Three of the constraints consist in checking that the components of momentum balance; and since the tracks' momentum vectors are determined almost independently of any mass assignments (provided that the energy loss through the material of the detector is negligible), then if the momentum constraints are satisfied for one hypothesis, they will also be so for the other. Thus there is only one useful constraint – the energy one – for separating the two reactions.

Now the energy difference produced by changing the two pions in reaction (A 2.3) to the two kaons in reaction (A 2.3') is

$$\Delta E = (E_K - E_\pi)_1 + (E_K - E_\pi)_2$$

$$= \sqrt{(p_1^2 + m_K^2)} - \sqrt{(p_1^2 + m_\pi^2)} + \sqrt{(p_2^2 + m_K^2)} - \sqrt{(p_2^2 + m_\pi^2)}$$

$$\sim \tfrac{1}{2}(m_K^2 - m_\pi^2)\left(\frac{1}{p_1} + \frac{1}{p_2}\right), \tag{A 2.4}$$

where the last line is a good approximation for momenta p_1 and p_2 much larger than m_K. Thus this difference decreases as the momenta increase. In contrast, the measured energy imbalance in the reaction has an uncertainty which usually increases at least as fast as the momentum, and maybe as fast as the square of the momentum. At a large enough energy, this uncertainty is larger than (A 2.4), and hence it becomes very difficult to choose between these two hypotheses.

Distinguishing between two such reactions also becomes more difficult as the difference in the square of the masses becomes small. Thus determining whether a particular event contains pions or muons in the final state is almost impossible simply from measurements of the tracks' momentum vectors.

Example (vi)

Similar to Example (v) is the so-called permutation ambiguity. Here we are considering a specific 4-constraint hypothesis (for example, reaction A 2.3)) but we are uncertain about which of the three positively charged tracks is the π^+. Thus it is necessary to test this single reaction three times, each time choosing a different track as the π^+. For each of these, we have four constraints, but the three momentum constraints are essentially the same for each possibility. Thus, as before, we have only the one energy constraint which is useful in distinguishing among the various permutation ambiguities. Indeed when the magnitudes of the momenta of two of the tracks are equal then interchanging the mass assignments of these two tracks produces no change in the energy balance equation, and hence these two permutations of the appropriate hypothesis will have identical probabilities. Of course, two momenta are never identical, but they can be equal within errors; again the effect becomes more serious as the momenta of the tracks increase.

Which of the possible ambiguities is indeed correct does not affect the total number of events accepted as examples of that particular reaction, but some kinematic quantities (for example, the production angle of the π^+, or the mass of the $\pi^+\pi^-$ system) describing the final state do depend crucially on the particular choice of track assignments.

In order to distinguish between the types of ambiguities mentioned here or in Example (v) above, some form of particle identification may be more useful than kinematic fitting, especially at high energy.

Example (vii)

Hypotheses involving two or more undetected particles are worse than the zero-constraint category, in that we can use the conservation equations to calculate the missing energy and momentum components (and hence also the effective mass) of the unseen neutral system, but we have no equality to check that the event is indeed consistent with such a reaction (see also the comments on Example (iv)). We cannot, however, calculate the momenta of each of the unseen particles separately; even for just two unseen particles, we would need six equations to do this, and we have only four. Thus, for example, in the reaction

$$pp \rightarrow p\pi^+ n\pi^\circ \tag{A 2.5}$$

in which the neutron and π° are undetected, we can calculate the momentum, energy and effective mass of the $n\pi^\circ$-system, and we can check to see whether it lies within the mass region of the Δ°-resonance, but we cannot calculate the decay angle of the Δ° since this requires a knowledge of the neutron and π° momenta separately. Similarly, the effective mass of the $\pi^+\pi^\circ$-system is unobtainable.

Since the checks that we perform are only inequalities, such reactions are likely to be highly ambiguous.

The above example can in principle be upgraded to the completely calculable zero constraint case if both the missing neutral particles are assumed to come from the decays of separate resonances of known masses. The two extra mass constraints then supplement the four kinematic ones to enable the six missing momentum components to be calculated. This is a similar effect to that mentioned in Example (x) for Σ°s.

Since the number of produced particles (including neutrals) increases as a function of energy, kinematic fitting of the production reaction is more useful at lower energies.

Example (viii)

We now consider the case of a neutral particle decaying into two seen charged particles (see Fig. A 2.1(c)). This could be an example of

$$\text{or} \quad \begin{array}{ll} \Lambda \rightarrow p\pi^- & \tag{A 2.6} \\ K^\circ \rightarrow \pi^+\pi^-. & \tag{A 2.7} \end{array}$$

If the production position of the neutral particle is known, then so is its direction (simply the line joining the production and decay vertices P and D of Fig. A 2.1(c)). The only unknown in checking either decay (A 2.6) or (A 2.7) is then the magnitude of the momentum of the neutral particle, and hence the test involves three constraints. As before, however, all but one of these involve momentum conservation. Thus we can regard the three constraints as consisting of two which check whether the neutral particle at D is produced at P, and one constraint which distinguishes the mass of the neutral particle.

If the production position of the neutral particle is unknown, then there are three unknowns associated with the neutral particle and hence only one constraint to check either reaction (A 2.6) or (A 2.7).

If the mass of the neutral particle is unknown, the number of constraints is reduced by one. Since it is the energy constraint which is lost, we can only *calculate* the mass of the neutral particle from its decay products, rather than *check* via the constraints that it is sensible. The constraint equations will simply ensure that momentum is conserved at the decay point. If, however, a production fit can also be performed (see Example (ix) below), then we can *check* that a consistent mass for the unknown particle is obtained there.

Example (ix)

For the two stage process such as depicted in Fig. A 2.1 (c), we can try to perform a kinematic fit to the whole event. A suitable hypothesis – one of many – would be

$$\mathrm{K^-p \to \overline{K}^\circ \pi^+ \pi^- n}; \quad \mathrm{\overline{K}^\circ \to \pi^+ \pi^-}. \tag{A 2.8}$$

We now have four energy-momentum conservation equations at each vertex P and D. From these eight, we must subtract the number of unknown quantities (three momentum components of the neutron, and the scalar momentum of the $\overline{\mathrm{K}}^\circ$) in order to obtain the total number of useful constraints $(4 \times 2 - [3+1] = 4)$ in the fitting reaction sequence (A 2.8). It is, however, more realistic to count the constraints at each vertex P and D separately, especially when we try to distinguish (A 2.8) from other possible reaction sequences e.g.

$$\mathrm{K^-p \to \overline{K}^\circ p \pi^- \pi^\circ}; \quad \mathrm{\overline{K}^\circ \to \pi^+ \pi^-}. \tag{A 2.9}$$

Then at D, we have three constraints that check whether the decay tracks are consistent with a $\overline{\mathrm{K}}^\circ$ produced at P and decaying at D (and hence we should be able to decide with a good degree of confidence whether this is so), but at P, we have only one constraint to distinguish the different production reactions, and hence maybe the event will be ambiguous between these possibilities.

If we test a reaction in which there are no unseen particles at production (apart from the one that subsequently decays) e.g.

$$\mathrm{K^-p \to \overline{K}^\circ p \pi^-}; \quad \mathrm{\overline{K}^\circ \to \pi^+ \pi^-}, \tag{A 2.10}$$

then we have seven constraints overall, with four of them at the production vertex. Hence we should be reasonably confident about whether to accept or reject this hypothesis for any particular event.

Example (x)

We sometimes wish to test whether a reaction involves a Σ° which decays electromagnetically:

$$\Sigma^\circ \to \Lambda\gamma; \quad \Lambda \to p\pi^- \qquad\qquad (A\ 2.11)$$

with a lifetime of $\sim 10^{-16}$ secs. Thus the distance a Σ° travels before it decays is so short compared with the detector's spatial resolution that although the Σ° of course has a momentum vector that we wish to determine, the production and decay vertices can be considered as coincident (see Fig. A 2.1 (d)). Then to check if an event is consistent with the production reaction

$$K^-p \to \Sigma^\circ \pi^+ \pi^-, \qquad\qquad (A\ 2.12)$$

followed by the Σ°-decay sequence (A 2.11), we test the hypothesis

$$K^-p \to \Lambda\pi^+\pi^-\gamma; \quad \Lambda \to p\pi^- \qquad\qquad (A\ 2.13)$$

which normally would result in one constraint at the production vertex P (cf. Example (ix)). But here we have the extra constraint that the effective mass of the $\Lambda^\circ\gamma$ system must equal the Σ° mass. Thus, in addition to the three constraints checking that the charged tracks at E are consistent with being the decay products of a Λ° coming from D (cf. Example (ix)), there are two constraints at P ensuring that the production reaction is consistent with (A 2.12).

A reaction with a Σ° and another neutral particle from the production vertex, e.g.

$$K^-p \to \Sigma^\circ\pi^+\pi^-\pi^\circ \qquad\qquad (A\ 2.13')$$

involves one more unknown than we have equations, and so cannot be tested or calculated.

Example (xi)

We now examine reactions involving a Ξ°, which decays according to the sequence

$$\Xi^\circ \to \Lambda\pi, \quad \Lambda \to p\pi^-. \qquad\qquad (A\ 2.14)$$

An example of such a reaction could be

$$K^-p \to \Xi^\circ K^+\pi^-,\qquad\qquad\qquad\qquad (A\ 2.15)$$

followed by the Ξ° decay sequence (A 2.14) (see Fig. A 2.1(*e*)). This differs from the Σ° example in that the Ξ° lifetime of $\sim 10^{-10}$ secs is such that it will typically travel several cm before it decays, i.e. its production and decay points P and D are distinct.

In fitting the charged tracks at E as the decay products of a Λ°, we have only one constraint since the point D is not measured and hence the Λ-direction is unknown (cf. Example (viii)). Similarly we can fit the measured track vectors at the vertex P to the production reaction (A 2.15) with one constraint, there being three unknowns associated with the Ξ°-momentum. Then we can fit the $\Xi^\circ \to \Lambda\pi^\circ$ decay at the point D also with one constraint since at this stage both the Ξ°- and the Λ-vectors are known from the previous fits. Finally we have a geometrical constraint arising from the fact that the Λ vector as defined from the vertex E and the Ξ°-vector defined from the vertex P must intersect (within errors) at some point in space. (Remember that two straight lines in three dimensions do not in general intersect.) There are also the inequalities that the intersection point must be such that the lengths of the Ξ°- and the Λ-tracks must be positive, i.e. that the point D is essentially between P and E, rather than before P or after E.

A reaction such as

$$K^-p \to \Xi^\circ K^+\pi^-\pi^\circ\qquad\qquad\qquad\qquad (A\ 2.16)$$

involves a second neutral particle at the production vertex, and hence cannot be fitted or even calculated.

Example (*xii*)

There also exist charged particles with short lifetimes, e.g.

$$\Sigma^- \to n\pi^-.\qquad\qquad\qquad\qquad (A\ 2.17)$$

The lifetime of the Σ^- is $\sim 10^{-10}$ secs, so it will typically travel a few cm between production and decay. This is often not far enough for its track to have sufficient curvature in a magnetic field for the Σ^--momentum to be measured. In performing the fit of the decay (A 2.17) to the measured tracks, we thus have four unknown quantities (the Σ^- scalar momentum and the neutron's three components of momentum). The four conservation equations allow us to calculate these four unknowns. In some cases, no real solution exists; the measured component of momentum of the π^- transverse to the Σ^- direction of motion may be larger than the maximum value allowed in Σ^- decay (which is simply the decay π^- momentum for a Σ^- at rest). In other circumstances, we obtain two possible solutions of the equations.

If the Σ^- momentum vector does happen to be measured (because it lives several lifetimes before it decays, or because our magnetic field is very strong, etc.), then the fit to the decay (A 2.17) will have one constraint.

A typical reaction involving a Σ^- is

$$K^-p \rightarrow \Sigma^-\pi^+\pi^+\pi^- \tag{A 2.18}$$

(see Fig. A 2.1(f)). Apart from the possible constraint at the Σ^- decay point, we will have four constraints at production, since after the decay fit,† the Σ^- momentum is determined and there are no other unmeasured quantities. Alternatively, for the reaction

$$K^-p \rightarrow \Sigma^-\pi^+\pi^+\pi^-\pi^\circ \tag{A 2.19}$$

the extra π° results in only one constraint for the fit to the production reaction.

There are, of course, other decay sequences, but these do not involve any concepts different from those discussed above.

Thus, for example, from a constraint point of view, the decay of the charmed meson

$$D^+ \rightarrow K^-\pi^+\pi^+ \tag{A 2.20}$$

is like that of the neutral Λ or K° (since the D^+'s momentum will almost certainly be unmeasured, and the decay products are all charged). Similarly the decay mode

$$D^\circ \rightarrow K^-\pi^+\pi^\circ \tag{A 2.21}$$

will resemble that of the Σ^- of unknown momentum, as in each case there is one neutral decay product.

Example (xiii)

Finally we consider briefly the case where the decay vertex is *not* seen. This could be because the Λ or K° decays outside our visual detector, and the decay products are detected and measured only later in a downstream

† In practice we would probably perform the better constrained fit at the production vertex before attempting the poorly constrained fit to the Σ^- decay.

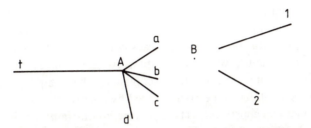

Fig. A 2.2. Diagrammatic representation of an event in which the beam track t interacts at A producing charged tracks (a–d) and a Λ, whose decay at point B occurs in a region of the apparatus where charged tracks are not detected. The proton and π^- from the decay of the Λ are observed downstream as tracks 1 and 2. For simplicity, we assume that there is no magnetic field in the region between where the Λ decays and where tracks 1 and 2 are first observed. From measurements on tracks 1 and 2, their directions are known, and perhaps also their momenta.

spectrometer. Alternatively, we could be producing short-lived particles (e.g. charmed or beauty particles, or τ-leptons) which decay before they reach the visible region of our detector.

To be specific, we consider the decay

$$\Lambda \to p\pi^-. \tag{A 2.22}$$

For simplicity, we assume that there is no magnetic field in the region between the production vertex (A of Fig. A 2.2) and where the proton and π^--tracks are first observed.†

The constraint counting differs from the case where the Λ decay point was visible (see Example (viii) above). Now, not only is the Λ momentum unknown but so is its direction; this reduces the number of constraints by two. We regain three constraints, however, by requiring that the Λ and the proton paths intersect in space (i.e. that the distance between lines defining these particles' trajectories is zero) and that the π^- track also passes through this point (two more constraints). Thus we have four constraints in all,‡ these being reduced to three or two in the event of one or both of the outgoing momenta being unknown.

† If a field were present in this region, the number of constraints would be unchanged, but the problem of finding a best fit solution would be complicated by the fact that the unseen portions of the proton and π^- tracks would become helices rather than straight lines.

‡ Since the geometric constraints are different in this case from in the situation where the Λ-decay point is visible, it is incorrect to deduce that the extra constraint (associated with the decay point B being unseen) makes the quality of the fit better.

With both outgoing tracks' momenta being unmeasured, the two constraints have a simple geometrical interpretation. In order for there to exist a decay point, the proton and π^- tracks must intersect in space (first geometrical constraint), while momentum conservation at the decay requires the Λ, proton and π^- to be coplanar, which implies that the production vertex A lies in the plane defined by the proton and π^- tracks (second geometrical constraint). These geometric constraints are independent of the particular 2-body decay mode.† In the event of the proton and/or π^- tracks' momenta being known, we obtain one or two extra kinematic constraints.

The four (or three or two) constraints that apply when the decay vertex is not seen can be thought of in an alternative way. We can regard the decay point B as having three unmeasured co-ordinates, which we wish to determine. For a given point B, there is only one unknown associated with the Λ, namely the magnitude of its momentum. To determine these unknowns we can use the four equations of energy–momentum conservation, together with the fact that the proton and the π^- tracks both must pass through B. Since the requirement that a point must lie on a line is two constraints, we have eight equations in all, and with only four unknowns, we have as before four constraints available for testing the event.

† The lack of discrimination among the various possible neutral decays is serious when it is important to obtain a complete and unbiassed sample of, say, Λ-decays. In some situations, however, this is not necessary. For example, if charmed particles decaying into $\Lambda\pi^+\pi^+\pi^-$ are being studied, they will appear in the mass spectrum of $\Lambda\pi^+\pi^+\pi^-$ as a sharp spike, and the presence of contamination in the Λ-sample will result simply in increasing the more-or-less smooth background below the charmed particle signal. Provided this extra source of background is not too large, it may be merely a nuisance.

Bibliography

A. Bjorck, 'Least squares methods in physics and engineering' (Lectures in Academic Training Programme at CERN), *CERN Report 81–16*.

R. Böck, 'Application of a generalised method of least squares for a kinematical analysis of tracks in bubble chambers', *CERN Report 60–30*.

S. Brandt, *Statistical and Computational Methods in Data Analysis* (North Holland Publishing Co., 1973); and 'Elements of probability and statistics', in 'Formulae and methods in experimental data evaluation' (ed. R. K. Böck *et al.* Published by the European Physical Society, 1984).

W. T. Eadie, D. Drijard, F. E. James, M. Roos and B. Sadoulet, *Statistical Methods in Experimental Physics'* (North Holland Publishing Co., 1971).

H. Eichinger and M. Regler, 'Review of track fitting in counter experiments', *CERN Report 81–06*.

A. Frodesen, O. Skeggestad and H. Tøfte, *Probability and Statistics in Particle Physics* (Bergen, 1979).

D. J. Hudson, 'Lectures on elementary statistics and probability'; and 'Maximum likelihood and least squares theory', *CERN Reports 63–29* and *64–18*.

F. James, 'Monte Carlo theory and practice', *Rep. Prog. Phys.*, **43** (1980), 1145; and 'Determining the statistical significance of experimental results', *CERN Report DD/81/02*.

S. L. Meyer, *Data Analysis for Scientists and Engineers* (Wiley, 1975).

J. Orear, 'Notes on statistics for physicists', Berkeley preprint UCRL-8417 (1958), unpublished.

R. R. Wilson, 'Monte Carlo study of shower production', *Phys. Rev.*, **86** (1952), 261.

INDEX